Physiology and Biochemistry of Plant Cell Walls

The first ever high-resolution image of a hydrated cell wall (Chinese water chestnut, *Eleocharis dulcis* L.) taken in ambient conditions using atomic force microscopy. The image is 1µm × 1µm (× 85 000) and clearly shows the layering of the cellulose microfibrils. Courtesy of A.R. Kirby, A. Patrick Gunning, Keith W. Waldron, Victor J. Morris and Annie Ng, 1995.

Physiology and Biochemistry of Plant Cell Walls

SECOND EDITION

C.T. Brett
Institute of Biomedical and Life Sciences,
University of Glasgow,
UK

and

K.W. Waldron
BBSRC Institute of Food Research,
Norwich,
UK

CHAPMAN & HALL
London · Weinheim · New York · Tokyo · Melbourne · Madras

Published by Chapman & Hall, 2–6 Boundary Row, London SE1 8HN, UK

Chapman & Hall, 2–6 Boundary Row, London SE1 8HN, UK

Chapman & Hall GmbH, Pappelallee 3, 69469 Weinheim, Germany

Chapman & Hall USA, 115 Fifth Avenue, New York NY 10003, USA

Chapman & Hall Japan, ITP-Japan, Kyowa Building, 3F, 2-2-1 Hirakawacho, Chiyoda-ku, Tokyo 102, Japan

Chapman & Hall Australia, 102 Dodds Street, South Melbourne, Victoria 3205, Australia

Chapman & Hall India, R. Seshadri, 32 Second Main Road, CIT East, Madras 600 035, India

First edition 1990

Second edition 1996

© 1990, 1996 C.T. Brett and K.W. Waldron

Typeset in 11/12pt Garamond by Saxon Graphics Ltd, Derby

Printed in Great Britain at the University Press, Cambridge

ISBN 0 412 58060 8

A catalogue record for this book is available from the British Library

∞ Printed on permanent acid-free text paper, manufactured with ANSI/NISO Z39.48-1992 and ANSI/NISO Z39.48-1984 (Permanence of Paper).

Contents

List of boxes and tables

Preface

We have sought in this book to present a series of portraits of the plant cell wall as it participates in various different aspects of the life of the plant cell. Hardly any event in the cell's life occurs without involving the wall in some way, and as a result the book covers almost every aspect of plant cell biology, albeit from a special point of view. In presenting the various portraits, we have tried to show how the biochemistry, physiology and fine structure combine to give a full picture. In many cases, however, cell wall research has not progressed far enough to give a complete picture, and numerous gaps remain.

In this second edition, we have introduced further aspects of cell walls, especially relating to human health and diet. In addition, we have updated all the chapters and greatly expanded the treatment of certain rapidly advancing topics, including wall architecture, extension, biotechnology and interactions with other organisms.

We are most grateful to Mike Black and John Chapman for inviting us to write this book and for their advice; to Clem Earle and Rachel Young for their encouragement and help; to Dr P.M. Dey for his helpful comments; to the many contributors of photographs and diagrams; to Ros Brett for taking more than her fair share of the parenting while writing was in progress; and, most especially, to Su Waldron for doing all the work on the word processor.

Acknowledgements

Figures

2.1a from Parker (1984) *Protoplasma* **120**, 224, Springer, Berlin; 2.1b, 3.14 a and b, 3.15, 3.16a, 3.17 a and b, 4.1a, 4.9a, 4.15 a and b, 6.3, 6.4b, 6.12 b and c, 6.13a, 7.3, 7.5, 8.3, 9.3 a and b, 10.2 a and b, 10.3, 10.5 a and b, 10.6 a and b, 10.7 a and b, 10.8, 10.9, kindly supplied by M. Parker; 2.2 kindly supplied by B. Wells and K. Roberts; 2.4 redrawn from Preston's (1974) modification of Meyer and Misch (1936) *Helv. Chim. Acta* **20**, 232–44; 2.6 kindly supplied by M. McCann; 2.12 redrawn from Selvendran (1985) *J. Cell Sci. Suppl.* **2**, 55–8, © (1985) The Company of Biologists Limited; 3.2 redrawn from Freudenberg and Neish (1968) in *Constitution and Biosynthesis of Lignin*, Springer, Berlin; 3.3 and 3.4 reproduced with permission from the *Ann. Rev. Plant Physiol.* 37, © (1986) by Annual Reviews Inc; 3.6a adapted from Morris *et al.* (1982) *J. Mol. Biol.* **155**, 507–16; 3.7 a and b redrawn from Lamport and Epstein (1983); 3.8 a, b, c and d and 3.11 kindly provided by M. McCann, P. Linstead and P. Knox; 3.13a from Parker (1979) *Can. J. Bot.* **57**, 2399–407; 3.13b adapted from Neville and Levy (1984) in *Biochemistry of Plant Cell Walls* (Society of Experimental Biology Seminar Series 28), Cambridge University Press, Cambridge; 4.3 redrawn from Feingold and Avigad (1980) in *The Biochemistry of Plants*, Academic Press; 4.4 a and b redrawn from Waldron and Brett (1984) in *Biochemistry of Plant Cell Walls* (Society of Experimental Biology Seminar Series 28), Cambridge University Press, Cambridge; 4.6 redrawn from Robinson *et al.* (1984) in *Biochemistry of Plant Cell Walls* (Society of Experimental Biology Seminar Series 28), Cambridge University Press, Cambridge; 4.7 redrawn from Higuchi *et al.* (1977) *Wood Sci. Technol.* **11**, 153–67; 3.10, 3.11 and 3.12 kindly supplied by Giddings *et al.*, reproduced from *J. Cell Biol.* (1980) **84**, 327–29, by copyright permission of the Rockerfeller University Press; 4.16 from Lloyd and Wells (1985) *J. Cell Sci.* **75**, 225–38, by copyright permission of The Company of Biologists Ltd; 5.2 kindly supplied by E. McRobbie; 5.3 adapted from Ward (1971) *Mechanical properties of solid polymers*, reproduced with copyright permission of John Wiley & Sons Ltd; 5.4 and 5.5 adapted from Richmond *et al.* (1980) *Plant Physiol.* **65**, 211–17; 5.8 redrawn from Cleland (1967) *Ann. N. Y. Acad. Sci.* **144**, 3–18; 5.9 redrawn from Cleland (1967) *Planta* **74**,

197–209; 5.14 a and b redrawn from Tietze-Hass and Dorffling (1977) *Planta* **135**, 149–54; 5.15 a and b redrawn from Moll and Jones (1981) *Planta* **152**, 442–9; 5.21 redrawn from Fry (1986) reproduced with permission from the *Ann. Rev. Plant Physiol.* **37**, © 1986 by Annual Reviews Inc.; 6.4a kindly supplied by M. Parker and J. Sargent; 6.6 b and c, 6.12d, redrawn from Weier *et al.* (1974) *Botany: An Introduction to Plant Biology* reproduced with copyright permission of John Wiley & Sons Ltd.; 6.7 a–d, 6.11 a–g, 6.13 b–e, 6.14 a and b, 8.1 a–c, from Gunning and Steer (1975) *Ultrastructure and The Biology of Plant Cells,* Edward Arnold, London, pictures kindly supplied by M.W. Steer; 6.10 redrawn from Robards (1976) in *Intercellular Communication in Plants: Studies on plasmodesmata* (eds B.E.S. Gunning and A.W. Robards) Springer, Berlin; 7.1 kindly supplied by B. Lewis; 7.2 from O'Connell *et al.* (1985) *Physiol. Plant Path.* **27**, 75–98; 7.4 from Collinge *et al.* (1987) *Plant Mol. Biol.* **8**, 405–18; 7.5 from O'Connel and Bailey (1986) in *Biology and Molecular Biology of Plant-pathogen Interactions*; J.A. Bailey (ed.) Springer, Berlin; 7.9 a–f from Robertson *et al.* (1985) *J. Cell Sci.* (S2) 317–31 by copyright permission of The Company of Biologists Limited; 8.5 and 8.6 kindly supplied by M.W. Steer; 8.7 redrawn from Reid and Bewley (1979) *Planta* **147**, 145–50; 9.2 a and b kindly supplied by M. Parker from Parker (1984) *Protoplasma* **120**, 233–41; 9.5 kindly supplied by Rentokil from C.R. Coggins, *Decay of Timber in Buildings* with copyright permission of Rentokil Ltd.

Abbreviations

AGP	arabinogalactan-protein
Ara	arabinose
ATPase	adenosine-triphosphatase
DCB	dichlorobenzonitrile
D_e	elastic deformation
DE	elastic compliance
DF	dietary fibre
D_p	plastic deformation
DP	plastic compliance
$d\pi$	osmotic pressure difference across plasma membrane
ER	endoplasmic reticulum
f	furanose
FTIR	Fourier-transform infra-red
Fuc	fucose
g	gravity
GA	gibberellic acid
Gal	galactose
GAX	glucuronoarabinoxylan
GDP	guanosine diphosphate
GalA	galacturonic acid
Glc	glucose
GlcA	glucuronic acid
h	hour
HPLC	high-performance liquid chromatography
HRGP	hydroxyproline-rich glycoproteins
Man	mannose
mRNA	messenger ribonucleic acid
NAA	naphthyl acetic acid
NMR	nuclear magnetic resonance
NSP	non-starch polysaccharides
p	pyranose
PAL	phenylalanine-ammonia lyase
PG	polygalacturonase
PME	pectin–methyl esterase
PIIF	protein inhibitor inducing factor
P_t	turgor pressure

P_{th}	threshold turgor pressure
RG I	rhamnogalacturonan I
RG II	rhamnogalacturonan II
Rha	rhamnose
SCFA	short-chain fatty acid
SLG	S-locus-specific glycoprotein
TEM	transmission electron microscopy
TGN	trans-Golgi network
T_m	maximum stress-relaxation time
T_o	minimum stress-relaxation time
UDP	uridine diphosphate
WHC	water-holding capacity
XET	xyloglucan endotransglycosylase

1 The role of the cell wall in the life of the plant

For the plant cell, the cell wall is a frontier zone, where the cell encounters the challenge of the outside world. For the plant biologist, the cell wall is a research frontier, at which the old concept of a dead, structureless box has given way to a new picture of a highly complex structure performing a great diversity of functions in the life of the plant.

The first cell wall functions that were recognized were those dependent on its rigidity, namely the provision of strength and shape to the cell. Such structural roles are clearly important, and in woody plants the cell wall is thickened to such a degree that over 95% of the dry weight of wood may be contributed by the walls of its constituent cells. Thus the wall is a key factor in the achievement of large size and structural strength in plants.

However, the cell wall does much more than provide strength and shape. Its very rigidity poses a problem: why does it not prevent cell growth in expanding tissues? It is now known that cell expansion is governed to a large extent by selective, closely controlled, reversible weakening of certain areas of the cell wall. Unless the pressure exerted by neighbouring cells is too great, this weakening permits slow cell expansion in one or more dimensions, without loss of the overall structural integrity of the wall. The mechanism of this selective wall weakening is only beginning to be understood; because the control of plant growth is of enormous importance in agriculture, this is an area of intense research activity at present.

The presence of a rigid or semi-rigid wall poses a further problem for the plant cell. The wall is formed of cross-linked macromolecules which may present an obstacle to the movement of large molecules into and out of the cell. This means that large proteins and nucleic acids may not be able to enter or leave the cell, irrespective of any permeability barrier imposed by the plasma membrane. Furthermore, the wall possesses a net negative charge, which means that the movement of positively charged molecules through it may be retarded. These constraints seem to have influenced the plant cell's choice of cell-signalling molecules:

most of them are small and either neutral or negatively charged, in contrast to the polypeptides and low-molecular-weight amines and steroids which are commonly used as hormones in animal systems. The role of the cell wall in transport through the plant is by no means an entirely negative one, though, since the cell wall provides a ready path for the movement of some materials parallel to the cell surface. The walls of neighbouring cells are in direct contact, and so the plant's cell walls combine to form a major transport pathway, the **apoplast**.

The exclusion of large molecules and supra-molecular assemblies from the cell wall may place some limitations on cell–cell communication, but any such disadvantages are probably far outweighed by the corresponding advantages. The wall protects the cell from the vast majority of potentially pathogenic organisms, since even the smallest of them, the viruses, cannot penetrate the wall. This defensive role is not simply a passive one: those microorganisms that secrete enzymes which might bore a hole through the wall may be further countered by the active deposition of lignin or callose in the wall in response to the pathogen's attack. Nor is the defensive role of the wall limited to the attacks of microorganisms; the deposition of lignin and silica in the wall renders some plant tissues unpalatable to most animals as well.

The active deposition of new cell wall materials in response to attack by pathogens is an example of cell–cell interaction, and the most recently discovered roles of the cell wall are concerned with this aspect of the plant's life. Pathogenic attack may bring about the degradation of cell wall polysaccharides, and it is thought that oligosaccharides produced during this degradative process may diffuse into neighbouring cells and trigger defensive reactions there. Oligosaccharides derived from cell wall polysaccharides may also have a role in controlling other events in the plant's life, including cell extension and cell differentiation.

Degradation of cell wall polymers occurs not only during pathogenic attack but also as part of normal cell growth and development. Parts of the wall are subject to metabolic turnover during vegetative growth – i.e. synthesis and breakdown occur at the same time. Cell wall breakdown increases dramatically at certain points in development, such as seed germination, fruit-ripening, abscission and senescence. These breakdown processes are under close metabolic control, both in location and timing, and they provide the developmental flexibility that prevents the wall becoming a strait-jacket.

The cell wall has a major influence on the life of plants with respect to their interactions with herbivores. Though it is not as digestible as, for instance, starch or protein, it nevertheless makes a major contribution to animal nutrition through the action of gut microorganisms. Furthermore, wall polysaccharides in the human diet are now thought to have important effects on human health, both in providing fibre in the diet and through their anti-tumour effects. After suitable industrial treatment, wall polysaccharides contribute further to the human diet through brewing and through the production of edible mushrooms. All these nutritional influences have contributed to the choice of certain plants for agriculture and hence to the survival of particular plant species and varieties in the human-dominated ecosystems of today.

Table 1.1 Wall functions

Function	Relevant chapters
Provision of mechanical strength	2 and 3
Maintenance of cell shape	2 and 3
Control of cell expansion	5
Control of intercellular transport	6
Protection against other organisms	7
Cell signalling	5,7 and 8
Storage of food reserves	8 and 9
Nutrition and health effects	10

The variety of roles that the cell wall fulfils goes far to explain the architectural complexity which is now being revealed in the wall by modern analytical techniques. This complexity means that cell wall structure, and hence also cell wall biosynthesis, must be under close control. The following chapters explore the physiology and biochemistry of the cell wall in the light of the multiple functions which the wall carries out. The major functions, and the chapters in which they are discussed, are given in Table 1.1.

Summary

The plant cell wall has many functions. These include: the provision of strength and shape to the cell and rigidity to the whole plant; the control of cell growth by selective weakening of the wall; protection from attack by pathogens and predators; participation in cell–cell communication; and influencing interactions with herbivores through nutritional and health effects. This multiplicity of functions may explain why the cell wall is such a complicted structure and it means that wall structure and biosynthesis must be closely controlled.

Further reading

Brett, C.T. (1983) Cell walls do not a prison make. *New Scientist*, **99**, 693–695.

2 The molecular components of the wall

2.1 The layered structure of the wall

The plant cell deposits its wall as a series of layers. The earliest layers are deposited at cell division, and the cell then lays down further material between the plasma membrane and the earlier layers. Thus the first-formed material is found at the point where the cell wall adjoins the cell wall of the neighbouring cell, and the latest-formed layers are closest to the plasma membrane (section 4.1). Three clear-cut, major layers can be identified by electron microscopy (Figure 2.1). The earliest-formed layer, found at the centre of the double wall formed by two adjacent cells, is called the **middle lamella**. It is derived from the cell plate, which is laid down at cell division. Since the cell plate is initially quite narrow and is often stretched during cell growth, the middle lamella is frequently extremely thin. It is often thickest at the cell corners, and can be detected in the electron microscope as a region which generally stains darker than neighbouring layers.

Once the cell plate is complete, the daughter cells proceed to deposit the next major layer, the **primary cell wall**. This layer continues to be deposited as long as the cell is growing in surface area, and the thickness of the primary wall is thus maintained approximately constant at about 0.1–1.0 μm, even though the older parts have been stretched and hence become thinner during cell growth. Primary walls have a rather similar appearance in most cell types, with the striking exception of transfer cells (section 6.3). However, under some staining conditions

Figure 2.1 The main layers in a cell wall. (a) TEM of a growing cell wall from a developing cotyledon of *Lupinus angustifolium* (L.) showing primary wall and middle lamella only. Bar, 1μm. (b) TEM of a TS of mature, non-extending cells of wheat leaf sclerenchyma showing well-developed secondary walls. Bar, 2μm. ML, middle lamella; PW, primary wall; SW, secondary wall; CT, cytoplasm. In (b), lignification may occur starting at the middle lamella and extending towards the plasmalemma.

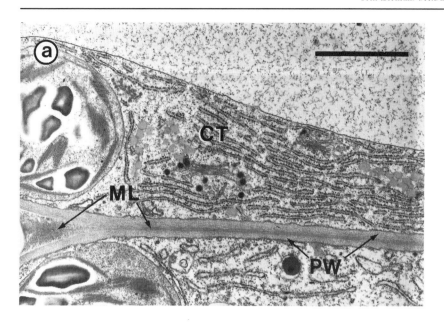

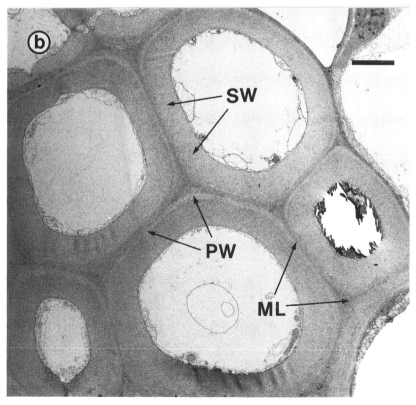

the microfibrils can be recognized, and in certain cell types the herring-bone pattern characteristic of helicoidal structures (section 3.5) can be seen.

Many cells limit themselves to the two layers mentioned above. Certain specialized cells, however, proceed to lay down a further, **secondary wall** at the onset of differentiation (Figure 2.1b). The secondary wall is variable in structure, both morphologically and chemically, and is indeed the most important diagnostic feature of some cell types. It is normally thicker than the primary wall, and may be laid down with varying thickness at different parts of the cell surface. The secondary wall thickenings which are formed in this way may produce a variety of patterns on the inner face of the wall, for instance in the spiral, annular, reticulate and scalariform thickenings of tracheid cell walls.

All the wall layers consist of two phases: a microfibrillar phase and a matrix phase (Table 2.1). The microfibrillar phase is distinguishable from the matrix phase by its high degree of crystallinity and its relatively homogeneous chemical composition. It is also readily visible in the electron microscope.

Table 2.1 Wall components

Phase	Components	
microfibrillar	cellulose (β1,4-glucan)	
matrix*	pectins	rhamnogalacturonan I
		arabinan
		galactan
		arabinogalactan I
		homogalacturonan
		rhamnogalacturonan II
	hemicelluloses	xylan
		glucomannan
		mannan
		galactomannan
		glucuronomannan
		xyloglucan
		callose (β1,3-glucan)
		β1,3-,β1,4-glucan
		arabinogalactan II
	proteins	extensin
		arabinogalactan-proteins
		others, including enzymes
	phenolics	lignin
		ferulic acid
		others, e.g. coumaric acid, truxillic acid

* NB Not all these matrix components are found in all cell walls.

2.2 The microfibrillar phase: cellulose

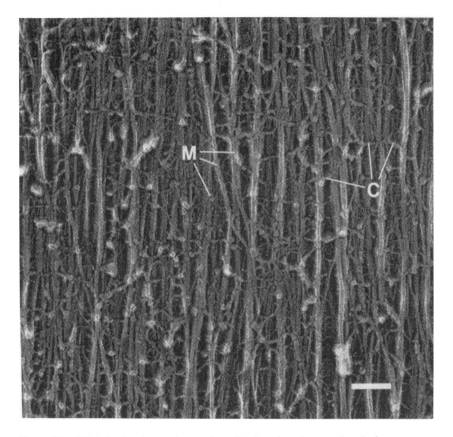

Figure 2.2   TEM of a fast freeze deep-etch replication of a primary cell wall of suspension-cultured carrot showing the microfibrillar structure in detail. Note the 'cross-links', C, between microfibrils (M), and the apparent space between them. Bar, 200 nm.

The microfibrillar phase of the wall is composed of extremely long, thin structures called **microfibrils** (Figure 2.2). They have a fairly uniform width, about 10 nm in mature higher plant cells and 20 nm in some algae, though there is some evidence that younger cells of higher plants (e.g. cambial cells) may have microfibrils as narrow as 3 nm. They are circular or oval in cross-section. The length of the microfibrils is uncertain, since it is seldom possible to identify their ends in electron micrographs with any confidence. The microfibrils are made up of cellulose molecules, which are aligned parallel to the long axis of the microfibril. About 30 to 100 cellulose molecules lie side-by-side at any one point along the microfibril. **Cellulose** is an unbranched β1,4-glucan, with a degree of **polymerization** (number of sugar residues per molecule) of up to at least 15 000. (For a brief overview of carbohydrate chemistry, see Box 2.1; for an explanation of polysaccharide linkage terminology, see Box 2.2; for examples of sugar structure, see Box 2.3.) There is some evidence for a lower degree of polymerization in primary walls than in secondary walls, and a rather precise chain length of around 15 000 glucose residues has been claimed for secondary wall cellulose of cotton. However, the chain length of such a large, insoluble molecule is difficult to measure with confidence, due to possible enzymic and mechanical degradation during analysis, and must be regarded as uncertain at present.

Box 2.1   Carbohydrates

Carbohydrates are polyhydroxy compounds with the empirical formula $C_x(H_2O)_y$. They vary in size, ranging from small molecules, e.g. monosaccharides (sugar residues), to very large ones, e.g. polysaccharides. Carbohydrates may be divided into three main classes:

1. Monosaccharides, such as glucose.
2. Oligosaccharides (up to approximately 10 monosaccharides joined by glycosidic linkages).
3. Polysaccharides (many monosaccharides joined by glycosidic linkages).

The most abundant polysaccharides are polymers of glucose:

(a) The storage polymer starch (1–4-linked α-D-Glc).
(b) The cell wall structural polymer cellulose (1–4-β-D-Glc).

Monosaccharides

These are the building blocks of oligosaccharides and polysaccharides, and have the empirical formula $(CH_2O)_n$, where n is equal to or greater than 3. Most common monosaccharides have an unbranched carbon skeleton. One of the carbon atoms has a carbonyl oxygen. If the carbon concerned is a terminal carbon atom, the sugar is known as an **aldose**. If it is elsewhere, the sugar is known as a **ketose**. All the remaining carbons have a hydroxyl group. The number of carbon atoms in monosaccharides can vary. The most simple would be the 3-carbon sugars glyceraldehyde (an aldotriose; Scheme 1) and dihydroxyacetone (a ketotriose; Scheme 2). The most abundant ketoses and aldoses contain 6 carbon atoms, the keto- and aldo-hexoses.

Stereoisomerism of monosaccharides

Except for dihydroxyacetone, all monosaccharides contain one or more asymmetric carbon atoms. They are, therefore, chiral molecules and will rotate plane-polarized light. The most simple example is glyceraldehyde which contains one asymmetric carbon atom. This allows the molecule to exist as two stereoisomers (Scheme 3), the D- and L-enantiomers. It should be noted that the symbols D and L refer only to the absolute configuration of the carbons, *not to the direction of rotation of plane-polarized light*. In larger sugars, the D and L configuration is determined by the orientation of the -OH group bound to the last asymmetric carbon atom in the carbon chain, i.e. the carbon furthest from the carbonyl group. Scheme 4 shows the projection formulas of the D-aldoses containing up to 6 carbons. In each case, the orientation of the -OH group in the last asymmetric carbon atom is the same. As the carbon chain length increases, more asymmetric carbon atoms are observed, giving rise to further stereoisomers of a given enantiomer. L-aldoses would be mirror images of the D-aldoses. Similar series can be derived from dihydroxyacetone (the D- and L-ketoses). D- and L-enantiomers of a monosaccharide will exhibit identical chemical and physical properties (apart from chirality and chemical interaction with other chiral compounds). However, stereoisomers of each enantiomeric form will possess different physical as well as chemical properties.

Ring structure of monosaccharides

Aldehydes and ketones can react with alcohols to form hemiacetals and hemiketals. Sugars, which contain either aldehydic or ketone groups in combination with hydroxyl groups, thus have the potential to form cyclic hemiacetals or hemiketals (see β-D-glucopyranose in Scheme 5).

Hemiacetal and hemiketal rings have either 5 (furanose) or 6 (pyranose) members. Rings with more than 6 or fewer than 5 members are generally unstable. The formation of the hemiacetal and hemiketal ring gives rise to a new chiral centre at the carbon which was attached to the carbonyl oxygen. Hence, the hemiacetal and hemiketal rings formed can be in two forms known as the α and β anomers (see the following section and Scheme 6).

Mutarotation

The open chain forms of sugars thus have the potential to form either furanose or pyranose rings, each with α- and β-anomers. This, and the fact that ring structures and open chain forms can exist in equilibrium when in solution (opening and closing constantly), results in the interconversion shown for β-D-glucopyranose in Scheme 6. The different forms have different thermodynamic stabilities, resulting in greater abundance of some forms compared with others.

Conformation of monosaccharides

Howarth projection formulas are the most convenient way of depicting a sugar residue. However, whilst they indicate the ring forms of monosaccharides, they may mislead the reader into thinking of rings as planar. *This is not the case.* Due to the tetrahedral arrangement of bonds about the carbon atoms, the pyranose form of monosaccharides may exist in two conformations known as the chair and boat conformations (Scheme 7). The substituent hydroxyls and hydrogen groups may be arranged either axially or equatorially (Scheme 8). The most thermodynamically stable position for hydroxyl groups is the equatorial orientation in the chair form; β-D-glucopyranose, therefore, is a very thermodynamically stable monosaccharide with all of its O-containing substituent groups in the equatorial plane (Scheme 9). Equatorially orientated OH groups are, generally, more easily esterified than axial groups.

Glycosidic linkages

Glycosidic linkages are formed as a result of a chemical or enzymically catalysed reaction between the anomeric carbon and an OH group (Scheme 10). The anomeric configuration of the resultant glycoside will depend on the configuration of the anomeric carbon. If the OH group is attached to a carbon of another sugar, the resulting dimer is known as a disaccharide. Addition of further sugars results in the formation of trisaccharides, tetrasaccharides, etc.

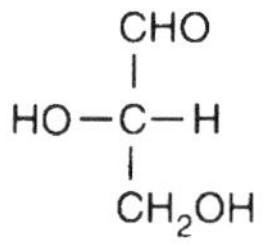

$$
\begin{array}{c}
\text{CHO} \\
| \\
\text{HO}-\text{C}-\text{H} \\
| \\
\text{CH}_2\text{OH}
\end{array}
$$

Scheme 1 Projection formula of glyceraldehyde

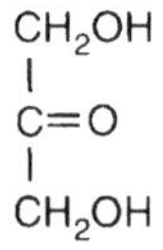

$$
\begin{array}{c}
\text{CH}_2\text{OH} \\
| \\
\text{C}=\text{O} \\
| \\
\text{CH}_2\text{OH}
\end{array}
$$

Scheme 2 Projection formula of dihydroxyacetone

```
        CHO                          CHO
         ⋮                            ⋮
   H►C◄OH                      HO►C◄H
         ⋮                            ⋮
       CH₂OH                        CH₂OH
```

Scheme 3 Perspective formulas of D-glyceraldehyde (left) and L-glyceraldehyde (right)

```
                        CHO
                        HCOH
                        CH₂OH
                   D-glyceraldehyde

         CHO                              CHO
        HOCH                             HCOH
        HCOH                             HCOH
        CH₂OH                            CH₂OH
       D-threose                       D-erythrose

   CHO          CHO          CHO          CHO
   HCOH        HOCH         HOCH          HCOH
   HOCH        HOCH         HCOH          HCOH
   HCOH        HCOH         HCOH          HCOH
   CH₂OH       CH₂OH        CH₂OH         CH₂OH
  D-xylose    D-lyxose     D–arabinose   D-ribose

 CHO   CHO   CHO   CHO   CHO   CHO   CHO   CHO
 HCOH  HOCH  HCOH  HOCH  HOCH  HCOH  HCOH  HOCH
 HCOH  HCOH  HOCH  HOCH  HOCH  HOCH  HCOH  HCOH
 HOCH  HOCH  HOCH  HOCH  HCOH  HCOH  HCOH  HCOH
 HCOH  HCOH  HCOH  HCOH  HCOH  HCOH  HCOH  HCOH
 CH₂OH CH₂OH CH₂OH CH₂OH CH₂OH CH₂OH CH₂OH CH₂OH
             CH₂OH
D-gulose  D-idose D–galactose D–talose D–mannose D–glucose  D–allose D–altrose
```

Scheme 4 D-aldose family, from 3 to 6 carbon atoms

```
          ┌──────────┐
    HO ─ C ─ H        │
         │            │
     H ─ C ─ OH       │
         │            │
    HO ─ C ─ H    O   │
         │            │
     H ─ C ─ OH       │
         │            │
     H ─ C ───────────┘
         │
       CH₂OH
```

Scheme 5 Projection formula of β-D-glucopyranose

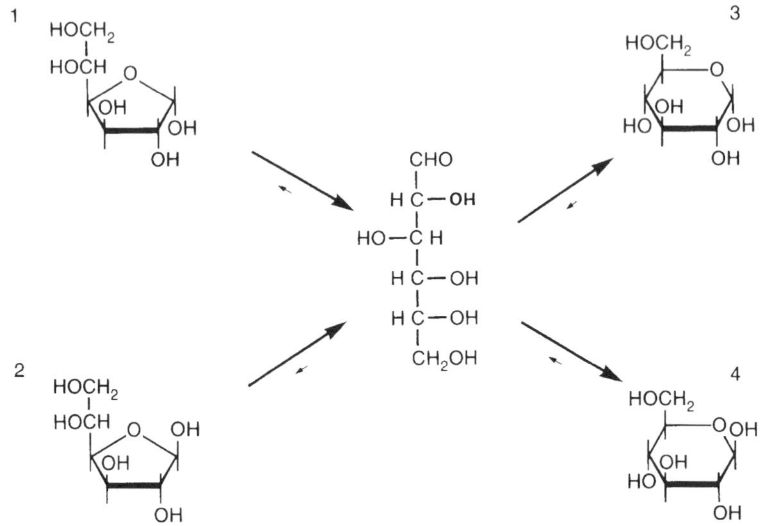

Scheme 6 Howarth projection formulas of D-glucopyranose, showing unstable furanose forms and stable pyranose forms. 1: α-D-glucofuranose (<1% total), 2: β-D-glucofuranose (<1% total); 3: α-D glucopyranose (36% total); 4: β-D-glucopyranose (64% total)

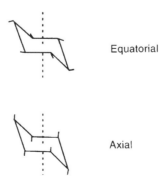

Scheme 7 Chair and boat forms of hexoses

Scheme 8 Axial and equatorial arrangement of side groups

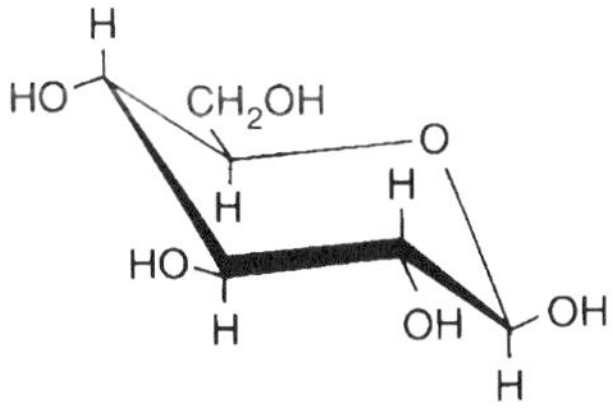

Scheme 9 β-D-glucopyranose in the chair form

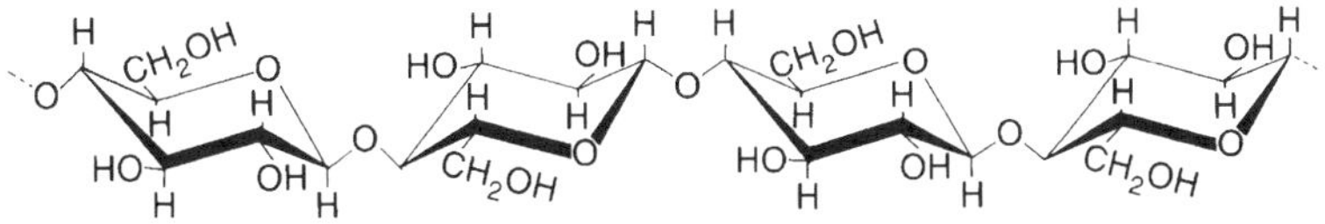

Scheme 10 (1–4)-linked β-D-glucopyranose

Box 2.2 Polysaccharide linkages

The monosaccharides in a polysaccharide are linked together by o-glycosidic bonds, in which the glycosidic carbon atom is linked through an oxygen atom to a carbon atom on another monosaccharide residue. The glycosidic carbon atom is the one that is linked to two oxygen atoms in the ring form of the monosaccharide; in cell wall polysaccharides, it is almost always carbon atom 1, known as C1.

The linkage pattern in a polysaccharide can be described in two ways. One way is to indicate the atoms at each end of one glycosidic bond. The bonds linking glucose residues in cellulose are thus 'β(1–4) bonds', the 'β' referring to the anomeric configuration of the glycosidic carbon atom.

Alternatively, it is possible to specify those carbon atoms on one particular residue which participate in glycosidic bonds. In cellulose, for instance, carbon atoms 1 and 4 on any individual glucose residue are involved in glycosidic bonds, and hence cellulose can be described as a 'β1,4–glucan', and each individual residue can be described as 'β1,4–linked'. It is important to note that this way of describing the structure uses a comma, but no brackets.

In cellulose, it makes little difference which terminology is involved. However, in more complicated polysaccharides the distinction becomes more important. For instance, in rhamnogalacturonan I, the backbone consists of α1,2–linked rhamnose and α1,4–linked galacturonic acid. The bond linking the C1 of rhamnose to galacturonic acid is α(1–4), while the bond linking the C1 of galacturonic acid to rhamnose is α(1–2) (see Figure 2.7a).

The difference in terminology becomes even more marked with branched polysaccharides. The arabinan shown in Figure 2.8 is linked by three kinds of bonds – $\alpha(1–2)$, $\alpha(1–3)$ and $\alpha(1–5)$ – but contains four types of arabinose residues – α1-linked, α1,5-linked, α1,2,5-linked and α1,3,5-linked.

Box 2.3 Structures of sugars commonly found in the plant cell wall

Box 2.3 continued

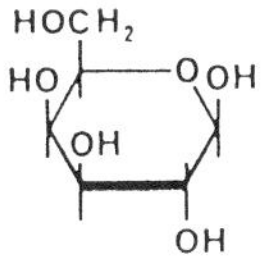

HOCH₂ ... β-D-galactose

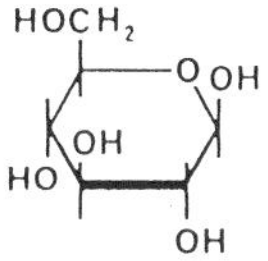

HOCH₂ ... β-D-glucose

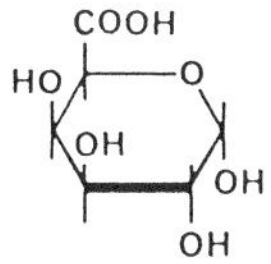

COOH ... α-D-galacturonic acid

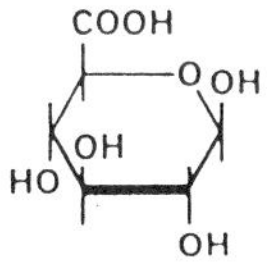

COOH ... β-D-glucuronic acid

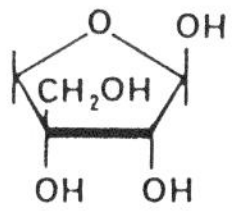

... β-D-apiose

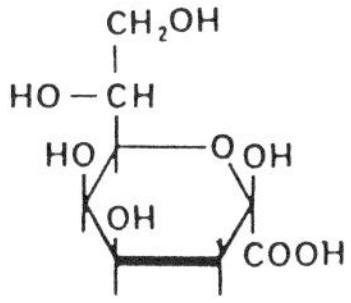

... ketodeoxyoctulosonic acid (KDO)

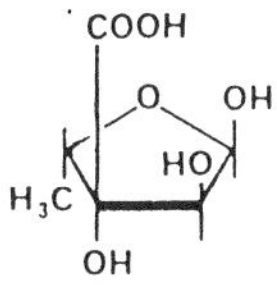

... β-L-aceric acid

The cellulose chains are held in a crystalline or paracrystalline lattice within the microfibril, giving rise to a structure of considerable tensile strength. The lattice is stabilized by both intramolecular and intermolecular hydrogen bonds. These are thought to form between the ring oxygen of one glucose residue and the C3 hydroxyl hydrogen of an adjacent residue within the molecular chain, and also between hydroxyl and oxygen atoms on neighbouring chains. There is still some uncertainty as to whether neighbouring molecules are parallel or antiparallel (Figure 2.3); the X-ray diffraction pattern is difficult to interpret in this respect, and only for one species (an alga) has definite evidence been obtained. In this case, the chains appear to be parallel.

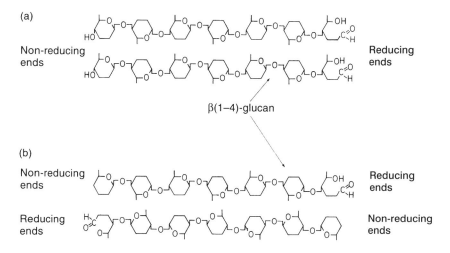

Figure 2.3   Schematic diagram illustrating (a) parallel and (b) antiparallel configurations of the glucan chains in the crystalline regions of cellulose. Note that in the antiparallel configuration, adjacent chains will be synthesized in opposite directions.

The internal crystal structure described above is known as **cellulose I**. Three further crystal structures are known (cellulose II, III and IV), which can be obtained from cellulose I by mechanical or thermal treatment. However, cellulose I is the only form commonly found in nature, even though it is metastable with respect to cellulose II, which is the most thermodynamically stable form.

It is clear from X-ray diffraction and chemical studies that the bulk of the microfibril is made up of crystalline β1,4-glucan (Figure 2.4). However, there is evidence that some degree of structural inhomogeneity may exist within the microfibril. First, the α-cellulose fraction of cell walls (obtained after solubilization and removal of lignin, pectin and hemicellulose) almost always contains a small amount of some sugars

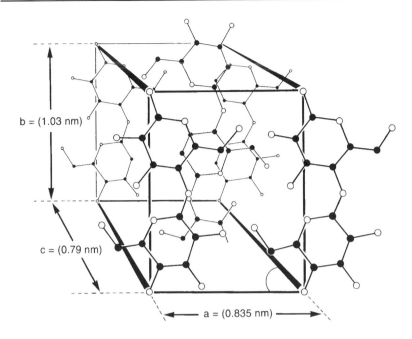

Figure 2.4   Diagram of the unit cell of cellulose as interpreted by Meyer and Misch (1936) and modified by Preston (1974). Note that in this model, the glucan chains are in an antiparallel configuration. ●, carbon, ○, oxygen.

other than glucose, usually mannose and xylose. Secondly, the X-ray diffraction pattern suggests a structure rather thinner than that seen by electron microscopy, indicating that the microfibril may consist of a crystalline core surrounded by a rather less crystalline surface region, or possibly a number of crystalline 'elementary fibrils' lying side by side. Thirdly, when negative stains are used to show up microfibrils in the electron microscope, there appear to be occasional points along the microfibril at which the heavy metal stain can penetrate into the centre of the microfibril, suggesting that less crystalline regions occasionally interrupt the central, crystalline core. These observations suggest a structure such as that shown in Figure 2.5, in which the non-glucose residues would be expected to be found in the less crystalline regions of the microfibril. Such a structure is supported by experiments in which the microfibril is disrupted by strong acid. This treatment gives rise to short rodlets of pure, fully crystalline $\beta1,4$-glucan, which may correspond to the fully crystalline regions of the microfibrillar structure shown in Figure 2.5. The mannose and xylose residues found in the less crystalline regions might be present as distinct mannans and xylans, or as heteropolymers together with glucose (possibly similar to the glucomannans and xyloglucans found in the matrix).

(a)

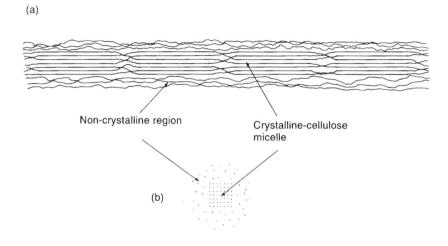

Non-crystalline region

Crystalline-cellulose micelle

(b)

Figure 2.5 Schematic diagrams of (a) LS and (b) TS through cellulose microfibrils illustrating the crystalline (micellar) and non-crystalline regions. The non-crystalline regions contain both cellulosic and non-cellulosic polysaccharides.

Microfibril structure is remarkably uniform throughout higher plants. Many algae also contain cellulose microfibrils. Certain algal taxa, however, contain microfibrils which, though similar in appearance in the electron microscope, are composed of long chains of β1,3-xylan or β1,4-mannan. In the fungi, microfibrils of chitin (poly-β1,4GlcNAc) take the place of cellulose.

2.3 The matrix phase: introduction

The non-crystalline phase of the cell wall is called the **wall matrix**. It appears relatively featureless in the electron microscope, but in chemical terms the matrix is extremely complex. It consists of a variety of polysaccharides, proteins and phenolic compounds, and the composition varies in different parts of the wall, in different types of cell, in different species and probably also at different stages of the cell cycle. These variations in composition include those not only in the proportions of the polymers present but also in the detailed structure of each polymer. None of the components are easy to analyse chemically and hence we are still some way from obtaining a complete picture of the structure of the matrix, even for the most intensely studied cell types. The following sections summarize our present knowledge.

Box 2.4 Procedure for the preparation of cell wall material from fresh tissue

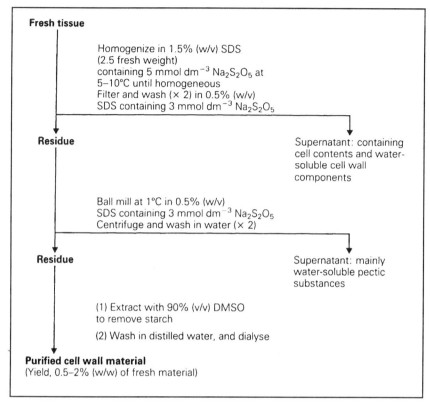

Fresh tissue

Homogenize in 1.5% (w/v) SDS
(2.5 fresh weight)
containing 5 mmol dm^{-3} Na$_2$S$_2$O$_5$ at
5–10°C until homogeneous
Filter and wash (× 2) in 0.5% (w/v)
SDS containing 3 mmol dm^{-3} Na$_2$S$_2$O$_5$

Residue

Supernatant: containing cell contents and water-soluble cell wall components

Ball mill at 1°C in 0.5% (w/v)
SDS containing 3 mmol dm^{-3} Na$_2$S$_2$O$_5$
Centrifuge and wash in water (× 2)

Residue

Supernatant: mainly water-soluble pectic substances

(1) Extract with 90% (v/v) DMSO to remove starch

(2) Wash in distilled water, and dialyse

Purified cell wall material
(Yield, 0.5–2% (w/w) of fresh material)

The main experimental approach used in studying the wall matrix has involved purification of the cell wall (Box 2.4) followed by extraction of wall components, either as impure fractions or as more highly purified components, followed by physical and chemical studies of the extracted material. This approach has yielded an enormous amount of information, but it has also imposed some limitations in our view of the wall.

The main constraint is that a certain number of bonds must be broken in order to extract components from the wall. If any of these bonds are covalent ones, then the resulting picture of the covalent structure of the components will be distorted. This drawback can in principle be overcome by comparing different extraction methods and especially by comparing chemical and enzymic extraction methods. A modern procedure for cell wall extraction, which minimizes polymer degradation by β-elimination (Selvendran, 1985), is given in Box 2.5 and 2.6 (see also Box 2.7).

A second limitation is that the majority of the matrix components are heterogeneous with respect to the details of their primary structure and molecular size. Hence extraction and purification of the components

Box 2.5 Extraction of pectic polysaccharides from cell wall material

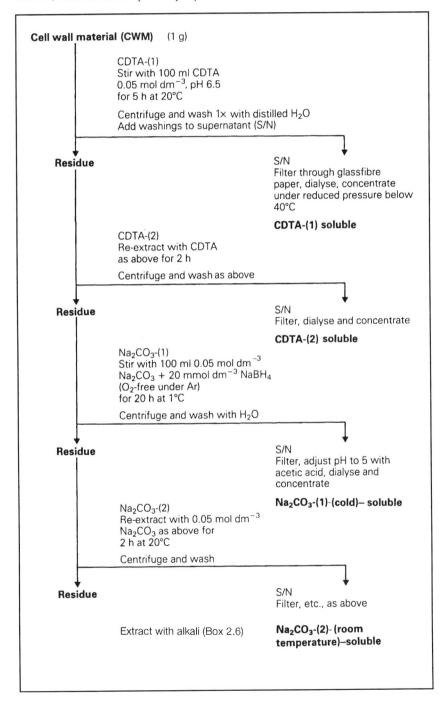

Box 2.6 Extraction of hemicelluloses from depectinated cell wall material

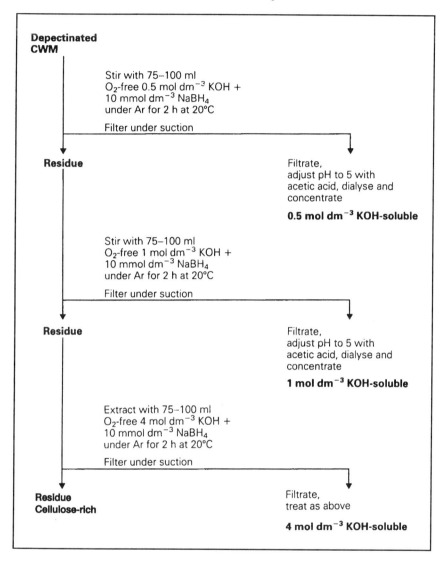

may result in only partial recovery of a particular type of molecule, and the recovered material may not be representative of the whole. This will lead to a distorted picture of the overall structure of the component.

In spite of these limitations, much has been learnt using the powerful tools that are now available for the analysis of extracted polymers. Polysaccharide sequencing methods make use of gas chromatography, HPLC and mass spectrometry, in combination with selective hydrolysis using purified enzymes. The proteins are studied by amino acid

Box 2.7   Units: conversion factors

The units used in this book are those most commonly used in the relevant literature. This box gives some translations for those more familiar with other conventions.

Length

$$1 \text{ nm} = 10^{-9} \text{ m}$$
$$1 \text{ μm} = 10^{-6} \text{ m}$$
$$1 \text{ mm} = 10^{-3} \text{ m}$$
$$1 \text{ cm} = 10^{-2} \text{ m}$$

Pressure

$$1 \text{ atmosphere} = 1.013 \text{ bar}$$
$$= 1.013 \times 10^5 \text{ Pascals (Pa)}$$
$$= 1.013 \times 10^5 \text{ Newtons metre}^{-2} \text{ (N m}^{-2})$$

Concentration

$$1 \text{ mol dm}^{-3} = 1 \text{ Molar (M) (mol l}^{-1})$$
$$= 10^{-3} \text{ mol m}^{-3}$$

sequencing and by DNA sequencing of the corresponding genes, while their secondary structures can be analysed spectroscopically. Only lignin remains largely unapproachable due to its highly irregular, three-dimensional linkage pattern (Chapter 3).

Few techniques are available for the analysis of the intact wall matrix. Some important information can be obtained from the overall sugar composition, from analysis of all the sugar linkages following methylation (Box 2.8), and from the overall amino acid composition. Spectroscopic techniques, especially NMR (Gidley, 1992), Raman and Fourier transform infra-red spectroscopy (FTIR) (Figure 2.6; McCann *et al*, 1992) are also beginning to yield information, whilst the distribution of matrix polymers within the wall has been investigated by electron microscopy using gold-labelled antibodies, lectins and wall-degrading enzymes. Atomic force microscopy is a novel technique which promises to yield much information concerning the fully hydrated cell wall (see Frontispiece, p. ii). Further, new techniques for the analysis of the intact matrix are eagerly awaited.

Box 2.8   Determination of sugar linkage in polysaccharides by methylation analysis

Methylation analysis gives information as to the position of substituted carbons (carbons to which side chains, groups and other sugars, are attached) within a sugar residue of a polysaccharide. In essence, free hydroxyl groups are methylated, and the polysaccharide hydrolysed to give a mixture of free, partially methylated sugars. In these, hydroxyl groups of those carbons which were involved in linkages to substituents or other sugars are non-methylated. The sugars are then reduced to give partially methylated alditols, and then acetylated to give volatile partially methylated, alditol acetates (PMAA) – see scheme. These may be separated by gas chromatography, the retention times of the PMAAs being determined by their degree of methylation, and the positions of the methyl groups. Accuracy of identification may be improved by use of mass spectrometry, and the use of sodium borodeuteride in the reduction step which allows accurate identification of C1.

a)

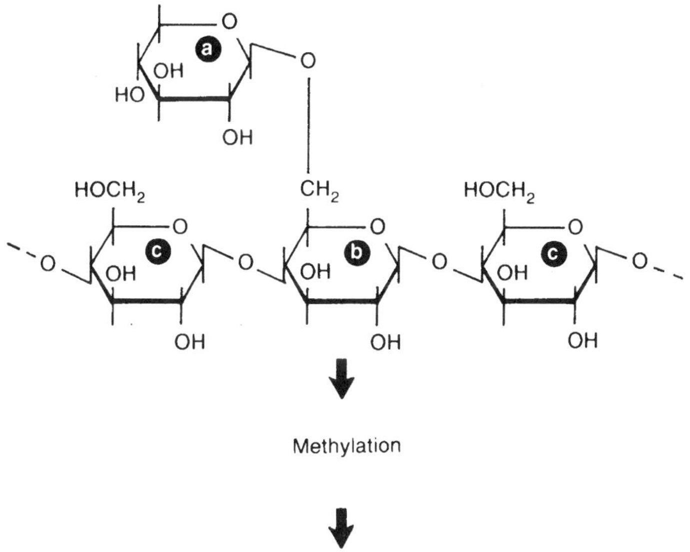

Methylation

Scheme iillustrating methylation analysis of residues within a xyloglucan polysaccharide. (a) Portion of xyloglucan showing three linkages common in this polysaccharide: a, b and c. (b) Same portion after methylation. Note that the free hydroxyls are now methylated. (c) Methylated monosaccharides, a, b and c, after release by acid hydrolysis. (d) Partially methylated alditols, after reduction with $NaBD_4$. (e) Partially methylated alditol acetates (PMAAs). Me, methyl group; Ac, acetyl group. Single hydrogens have been omitted.

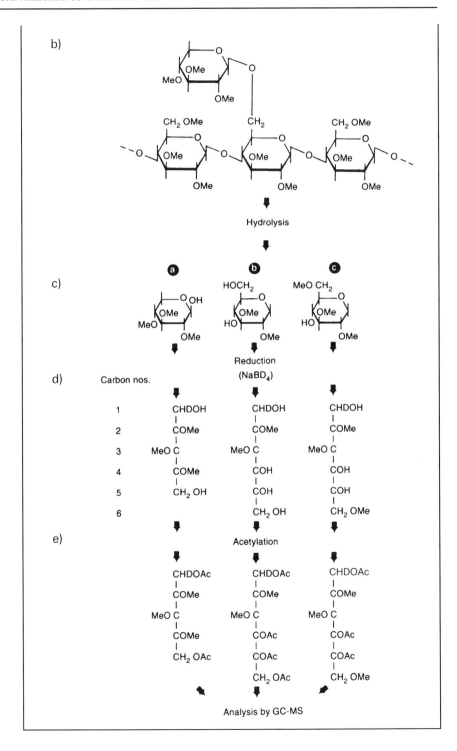

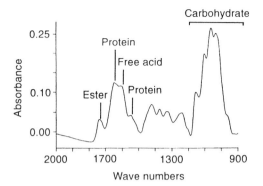

Figure 2.6 Fourier transform infra-red spectrum from an area of 50 x 50 μm of a single tomato cell wall fragment. Peak assignments from structure–frequency correlation charts show, for example, that this wall is rich in protein (absorbances at 1550 and 1650 cm^{-1}) and both esterified pectins (absorbance at 1740 cm^{-1}) and unesterified pectins (free acid absorbance at 1600 cm^{-1}).

2.4 Extraction and classification of matrix polysaccharides

The traditional classification of matrix polysaccharides has been based on the methods used for their extraction and, in some cases, reprecipitation. The wall, delignified if necessary, can be extracted either with a hot, aqueous solution of a chelating agent or with hot, dilute acid, yielding the fraction known as **pectin**. This fraction is generally rich in galacturonic acid, rhamnose, arabinose and galactose. After removal of pectin, the remainder of the matrix polysaccharides can be extracted using alkaline solution, yielding a fraction known as **hemicellulose**. This fraction can be further subdivided into hemicellulose A, which reprecipitates when the alkaline extract is neutralized, and hemicellulose B, which reprecipitates when ethanol is added to the neutralized extract to give a 70% ethanolic solution. The residue of the cell wall which remains insoluble after extraction with alkali is called **α-cellulose**, and contains the cellulose microfibrils.

This classification is convenient to use and has some physiological significance. However, it is unsatisfactory in that some polysaccharides are not clearly partitioned into one class. For instance, it is difficult to extract all the galacturonic-acid-containing material in the pectin fraction, and the α-cellulose fraction may contain traces either of 'pectin' (i.e. material containing galacturonic acid) or of hemicellulosic polysaccharides such as glucomannan and xyloglucan (section 2.2). For this reason some workers prefer not to subdivide the matrix polysaccharides into pectin and hemicellulose. However, the subdivision is useful in describing both the distribution of polysaccharides within the wall and the development of the wall, and hence it will be used in this book.

Further classification of individual polysaccharide types is made on the basis of the major sugars found in each polysaccharide.

2.5 Pectic polysaccharides

These are made up of a group of polysaccharides rich in galacturonic acid, rhamnose, arabinose and galactose. They are characteristic of the middle lamella and primary wall of dicotyledonous plants, and to a lesser extent of monocotyledonous plants. Under relatively mild conditions they are highly susceptible to β-eliminative degradation, so that some or all of the pectic polysaccharides described below may be covalently joined *in vivo*. They may also be linked covalently to phenols, cellulose and protein (section 3.2).

2.5.1 Rhamnogalacturonan I (RG I)

This polysaccharide is a major component of the middle lamella and primary cell wall of dicotyledonous plants, with the greatest concentration in the middle lamella. It contains a backbone of α1,4-linked galacturonic acid and α1,2-linked rhamnose. It has been studied most closely in suspension-cultured sycamore cells, from which it was extracted by partial digestion with endopolygalacturonase. In these cells the ratio of galacturonic acid to rhamnose is about 1 : 1, probably as a regular, alternating sequence. In the wall it may be covalently attached to homogalacturonan, i.e. to long stretches of galacturonic acid with little or no rhamnose (section 2.5.5). 'Hairy' regions of RG I, with many side-chains as described below, may alternate with 'smooth' regions of homogalacturonan to form very long molecules. Many of the galacturonic acid residues of RG I are methyl esterified, and some may contain acetyl groups esterified to their hydroxyl groups. The backbone is long, since even after extraction from the wall by endopolygalacturonase treatment, the degree of polymerization is around 2000. In tissues other than sycamore suspension cultures, the ratio of rhamnose to galacturonic acid in RG I may be lower and rhamnose may occur at less frequent intervals along the chain.

RG I contains a number of different side-chains, attached to the 4-position of rhamnose (Figure 2.7 a), These side-chains are composed principally of arabinose and galactose. Where arabinose is linked to rhamnose it is itself 1,5-linked, while galactose linked to rhamnose may be 1,4-linked. These side-chains are often quite large, and are similar in structure to the arabinan, galactan and arabinogalactan I molecules described below.

RG I from a range of tissues has been reported to contain additional, shorter side-chains (Figure 2.7). Since there is such a variety of possible side-chains, it may be that cells of different types or of different ages produce RG I molecules with different side-chains. As yet, however, there is no firm evidence for this idea. It is also possible that the rhamnogalacturonans of the middle lamella may differ in structure from those of the primary wall (section 2.5.7).

(a)

$\rightarrow$4)-α-GalA-(1$\rightarrow$2)-α-Rha-(1$\rightarrow$4)-α-GalA-(1$\rightarrow$2)-α-Rha(1$\rightarrow$

(b)

Ara-(1$\rightarrow$5)-Araol
Ara-Ara-Araol
Ara-Ara-Ara-Araol
Ara-Ara-Ara-Ara-Ara-Araol

Ara-(1$\rightarrow$4)-Rhaol
Ara-Ara-Rhaol
Fuc-Ara-Rhaol

Gal-Galol
Gal-Gal-Galol
Gal-Gal-Gal-Galol

β-D-Gal-(1$\rightarrow$4)-Rhaol
β-D-Gal-(1$\rightarrow$6)-β-Gal-(1$\rightarrow$4)-Rhaol

β-Gal-(1$\rightarrow$4)-Rhaol
β-Gal-(1$\rightarrow$4)-β-Gal-(1$\rightarrow$4)-Rhaol
α-Fuc-(1$\rightarrow$2)-β-Gal-(1$\rightarrow$4)-β-Gal-(1$\rightarrow$4)-Rhaol

β-Gal-(1$\rightarrow$4)-Rhaol
α-Ara-(1$\rightarrow$3)-β-Gal-(1$\rightarrow$4)-Rhaol
α-Ara-(1$\rightarrow$2)-α-Ara-(1$\rightarrow$3)-β-Gal-(1$\rightarrow$4)-Rhaol
Ara-(1$\rightarrow$5)-Ara-(1$\rightarrow$2)-Ara-(1$\rightarrow$3)-Gal-(1$\rightarrow$4)-Rhaol

β-Gal-(1$\rightarrow$4)-Rhaol
β-Gal-(1$\rightarrow$4)-β-Gal-(1$\rightarrow$4)-Rhaol
β-Gal-(1$\rightarrow$6)-β-Gal-(1$\rightarrow$4)-β-Gal-(1$\rightarrow$4)-Rhaol
Gal-Gal-Gal-Gal-Rhaol

Figure 2.7 Rhamnogalacturonan I. This polysaccharide consists of a backbone of α1,4-linked galacturonic acid and α1,2-linked rhamnose as shown in (a). Side-chains are attached to C4 of rhamnose. Examples of those side-chains that have been identified are given in (b); Araol, Galol, and Rhaol refer to arabinitol, galactitol and rhamnitol, respectively; these sugar alcohols are formed when the side-chains are cleaved from the backbone and indicate the point of attachment to the backbone.

2.5.2 Arabinan

This is a highly branched molecule containing a backbone of α1,5-linked arabinose and side-chains of single arabinose residues linked by α(1–2) or α(1–3) bonds to the main chain (Figure 2.8). Oligosaccharides of α1,5-linked arabinose may also be present as side-chains linked to the backbone in the same way, and other side-chains may be present in small amounts.

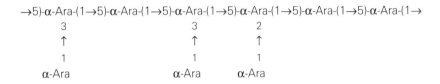

→5)-α-Ara-(1→5)-α-Ara-(1→5)-α-Ara-(1→5)-α-Ara-(1→5)-α-Ara-(1→5)-α-Ara-(1→
   3   3   2
   ↑   ↑   ↑
   1   1   1
α-Ara   α-Ara   α-Ara

Figure 2.8   Arabinan. A highly branched polysaccharide consisting of a backbone of α(1-5)-linked arabinose and side chains of α(1-2)- or α(1-3)-linked arabinose residues.

2.5.3 Galactan

Some primary cell walls contain β1,4-linked galactans, with little or no additional sugar material present in the molecule. In some cases a few of the galactose residues appear to be 1,6-linked.

2.5.4 Arabinogalactan I

These molecules contain a β1,4-linked galactan backbone to which short α1,5-linked arabinose side-chains are attached at C3 of galactose (Figure 2.9). They appear to exist either as independent molecules or as side-chains attached to rhamnogalacturonan I (section 2.5.1). Ferulic acid is found attached to some arabinose and galactose residues, and may be involved in cross-linking in the wall (section 3.2).

→4)-β-Gal-(1→4)-β-Gal-(1→4)-β-Gal-(1→4)-β-Gal-(1→4)-β-Gal-(1→4)-β-Gal-(1→
                         3   3
                         ↑   ↑
                         1   1
α-Ara-(1→5)-α-Ara-(1→5)-α-Ara   α-Ara

Figure 2.9   Arabinogalactan I. A β(1-4)-linked galactan backbone to which short α(1-5)-linked arabinose side chains are attached.

2.5.5 Homogalacturonans

Homogalacturonans are made up of α1,4-linked chains of galacturonic acid, which may be partly methyl esterified (Figure 2.10). Other sugar residues are absent or present only in low quantities. The homogalacturonans may be covalently attached to rhamnogalacturonan I or to other polysaccharides *in vivo*, since their extraction in a pure form requires the use of endopolygalacturonase or strong alkali. They are found in considerable quantities in some fruits and are also present in the primary walls of suspension-cultured dicotyledonous cells. Homogalacturonan with a low degree of methyl esterification may be referred to as '**pectic acid**', while more highly methylated molecules may be referred to as '**pectinic acid**', or simply as 'pectin'. This terminology can also apply to molecules containing covalently linked homogalacturonan and rhamnogalacturonan.

$$\rightarrow 4)\text{-}\alpha\text{-GalA-}(1\rightarrow4)\text{-}\alpha\text{-GalA-}(1\rightarrow4)\text{-}\alpha\text{-GalA-}(1\rightarrow4)\text{-}\alpha\text{-GalA-}(1\rightarrow4)\text{-}\alpha\text{-GalA-}(1\rightarrow$$

Figure 2.10   Homogalacturonan. This consists of α1,4–linked chains of galacturonic acid which may be partially esterified.

2.5.6 Rhamnogalacturonan II (RG II)

This material is found as a minor component of primary walls of suspension-cultured dicotyledonous cells. It is released from the cell walls by endopolygalacturonase action and may therefore be attached to rhamnogalacturonan I or homogalacturonan *in vivo*. As isolated, it has a degree of polymerization of about 60, and has a complex structure which includes galacturonic acid, rhamnose, arabinose and galactose in the ratio of 10 : 7 : 5 : 5, together with smaller amounts of some rare sugars such as aceric acid, apiose and 3-deoxy-manno-octulosonic acid (KDO). The arrangement of sugar residues is different from RG I, since the rhamnose residues are 1,3-, 1,3,4- or 1,2,3,4-linked or terminal. Partial acid hydrolysis of RG II has resulted in the liberation of the oligosaccharides shown in Figure 2.11.

2.5.7 Distribution of galacturonans in the middle lamella and primary wall

It has been suggested that the Rha- and GalA-containing pectins of the middle lamella may differ from those of the primary wall. The former appear to have fewer rhamnose residues, fewer and shorter branches, and a higher degree of esterification than the latter. In both cases, lengths of rhamnogalacturonan may alternate with lengths of homogalacturonan in the same molecule (Figure 2.12).

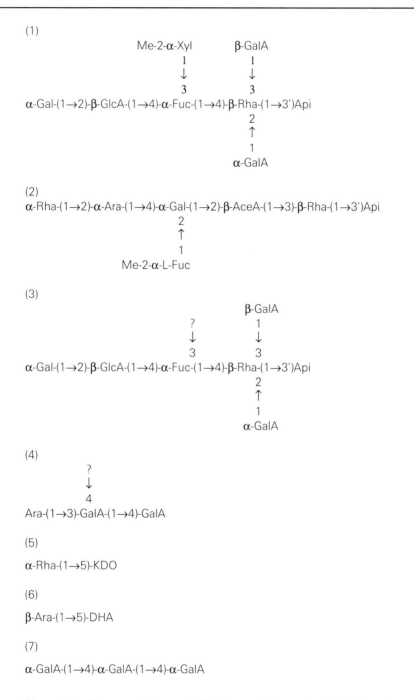

Figure 2.11   Rhamnogalacturonan II. Structures of oligosaccharides that have been released from RG II by partial acid hydrolysis.

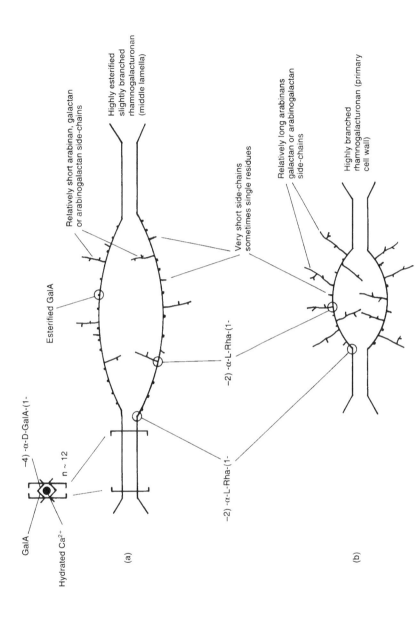

Figure 2.12 Diagrammatic representation of some structural features of pectins from the middle lamellae (a) and primary cell walls (b). Note that divalent calcium ions allow cross-linking between the homogalacturonan portions of the molecules resulting in an intricate matrix. See also Figure 3.5.

2.6 Hemicelluloses

The hemicelluloses are not extracted from the walls by hot water or by solutions of chelating agents or dilute acid. They can, however, be extracted with alkaline solutions, typically 1–4 mol dm⁻³ hydroxide. Once extracted, they may become more water-soluble. In many cases, the requirement for relatively strong alkali for their extraction from the wall is due to strong hydrogen-bonding between the hemicellulose and cellulose microfibrils. This probably applies to xylans, glucomannans and xyloglucans.

In contrast to the pectins, the hemicelluloses vary greatly in different cell types and in different species. In most cell types, one hemicellulose predominates, with others present in smaller amounts.

2.6.1. Xylans

These polysaccharides contain a backbone of β1,4-linked xylose residues. The backbone is substituted by α-linked 4-O-methylglucuronic acid on C2 of some xylose residues, by α-linked arabinose on C2 or C3 and by acetyl esters on C2 or C3 (Figure 2.13). Longer side-chains containing arabinose and other sugars are occasionally found. Different cell types have xylans which contain these substituents in different absolute and relative amounts. The primary walls of monocotyledonous plants contain, as a major hemicellulose, an arabinoxylan in which arabinose is the dominant side-chain. Secondary walls of monocotyledonous plants contain an arabinoxylan with rather more glucuronic acid. The primary walls of dicotyledonous plants have small amounts of a glucuronoarabinoxylan, containing both glucuronic acid and arabinose side-chains. The secondary walls of dicotyledonous plants contain glucuronoxylan, with only a low proportion of arabinose, as their major hemicellulose.

The distribution of side-chains in the glucuronoxylans of wood (i.e. in mature secondary walls) does not appear to follow a regular pattern.

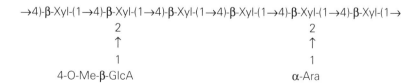

Figure 2.13 Xylan. A β1,4–linked chain of xylose residues which may be acetylated on either C2 or C3 or both. This backbone may be substituted by α–linked 4–0–methyl glucuronic acid on C2 of some xylose residues and by α–linked arabinose on C2 or C3. Other, longer side-chains containing xylose and arabinose, and even galactose, xylose and arabinose have also been reported.

There is, however, some clustering of glucuronic acid groups in certain regions of the xylan backbone of glucuronoxylan. In contrast, the glucuronoxylans of primary walls appear to have a regular spacing of glucuronic acid side-chains on every sixth xylose residue (Nishitani and Nevins, 1991). This is shown by the action of a specific glucuronoxylanase, which cleaves the xylan backbone close to the glucuronic acid side-chain. The distribution of arabinose side-chains along the backbone may follow a similar repeating pattern.

The xylans are capable of crystallization under certain conditions, though they are not thought to be crystalline in the cell wall. They may crystallize with either a two-fold or a three-fold screw axis, depending on the degree of acetylation.

The arabinoxylans of monocot primary walls are substituted by ferulic acid. These ferulic acid components may cross-link the arabinoxylans in the wall (section 3.2).

2.6.2. Glucomannans

Glucomannans form the major hemicellulose of the secondary cell walls of gymnosperms, as well as being a minor component of angiosperm secondary walls. They consist of a backbone of $\beta1,4$-linked glucose and mannose residues, in a ratio of about 1 : 3 in gymnosperms and 1 : 2 in angiosperms (Figure 2.14). There is no precise pattern in the sequence of glucose and mannose residues. In the case of gymnosperm glucomannan, single galactose residues are present as side-chains, in a molar ratio approximately equal to that of glucose, and these molecules are often referred to as galactoglucomannans. Some acetyl esters may also be present on the mannose hydroxyl groups.

→4)-β-Glc-(1→4)-β-Man-(1→4)-β-Man-(1→4)-β-Glc-(1→4)-β-Man-(1→4)-β-Man-(1→

Figure 2.14   Glucomannan. A polysaccharide of $\beta1,4$–linked glucose and mannose in a ratio of approximately 1 : 3.

2.6.3 Mannans and galactomannans

These molecules have been found only in the cell walls of some seed endosperms and some cotyledons, e.g. lupin, where they function as a food reserve. They contain a $\beta1,4$-linked mannose chain, and where galactose is present it is linked to mannose by an $\alpha(1$-$6)$ bond. The mannose : galactose ratio varies between species in legume seed endosperms. The distribution of the galactose side-chains is non-random, and is being investigated in relation to the mechanism of biosynthesis and subsequent processing of the polysaccharide (section 4.2). The mannans are able to form very hard, crystalline structures, and are laid down as microfibrils in some algae.

2.6.4. Glucuronomannan

This polysaccharide may be present in small amounts in a wide range of cell walls. It contains a backbone of α1,4-linked mannose residues and β1,2-linked glucuronic acid residues, perhaps in an alternating sequence. Side-chains include galactose linked β(1–6) to mannose, and arabinose linked (1–3) to mannose (Figure 2.15).

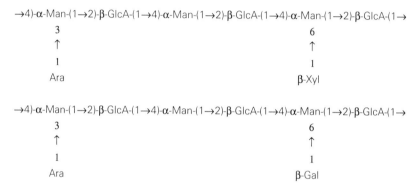

Figure 2.15   Predominant linkages and side-chains in glucuronomannan.

2.6.5. Xyloglucan

Xyloglucan is the principal hemicellulose of the primary walls of dicotyledonous plants. It consists of a backbone of β1,4-linked glucose residues, to the majority of which xylose residues are attached by α(1–6) bonds. Some xylose residues are further substituted by the di-saccharide Fuc α(1–2) Gal β(1–2), and occasionally by Ara (1–2) in some tissues. Acetyl esters may also be present, with a high proportion of the galactose residues carrying acetyl groups. In some tissues, at least, the xyloglucan is thought to be built up of alternating nonasaccharide and heptasaccharide units (Figure 2.16). In other tissues the regularity of structure is less marked, and the molecule is better thought of as a block copolymer. However, the less regular structure might arise by modification (e.g. by partial degradation within the wall) of what was ini-tially a regular structure.

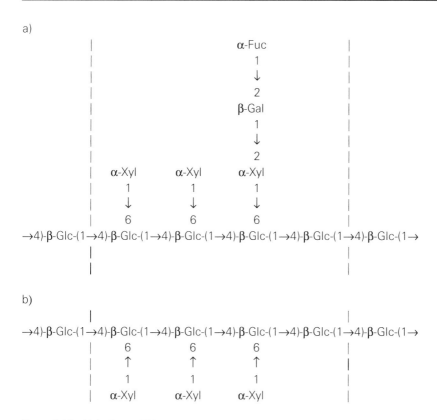

Figure 2.16 Xyloglucan. This polysaccharide consist of a backbone of β(1–4)–linked glucose residues to which xylose residues are attached by α(1–6) bonds. Other residues are often attached to the xylose side chains and the nonasaccharide (a) and heptasaccharide (b) repeat structures have been identified in xyloglucan cell-wall polysaccharides. Note that the fucose residue in (a) is not present in xyloglucan storage polysaccharides in seeds.

Xyloglucans are also found as storage polysaccharides in some seed endosperm cell walls (section 7.4). In these tissues, the structure generally lacks fucose. Primary walls of monocots contain small amounts of xyloglucan, with a lower xylose : glucose ratio than in dicot primary walls.

The conformation of xyloglucans has been predicted by computer modelling. A sequence of the heptasaccharides shown in Figure 2.16b is predicted to form a slightly twisted ribbon. However, the presence of the fuc–gal–xyl side-chain in the nonasaccharide is likely to straighten the chain, producing a conformation which should favour hydrogen-bonding to cellulose. The presence of arabinose side-chains is likely to prevent binding to cellulose (Levy *et al.*, 1991).

2.6.6. Callose

Callose is a general name for β1,3-glucan, which is found in a number of special situations in plant cell walls (Figure 2.17). The best-known situation is in phloem sieve tubes, where β1,3-glucan is found lining the pores of the sieve plate. Some of this callose may, however, be artefactual, since callose is formed by a variety of cells in response to wounding, and may be deposited on the sieve plates as plant material is harvested for study. Thus there appears to be an inverse correlation between the rapidity of tissue fixation and the amount of callose observed on the sieve plate by electron microscopy. Other situations in which callose is formed as a response to wounding include mechanically agitated suspension-cultured cells and plasmodesmata that are broken on cell separation. The β1,3-glucan that is observed on the surface of protoplasts in the early stages of wall regeneration may also be a reaction to wounding (i.e. to the loss of the original cell wall).

→3)-β-Glc-(1→3)-β-Glc-(1→3)-β-Glc-(1→3)-β-Glc-(1→3)-β-Glc-(1→

Figure 2.17 Callose. β1,3-linked glucan.

Callose is found in a number of other specialized tissues, including the pollen tube, the walls of some pollen, compression wood tracheids of gymnosperms and the secondary wall of cotton fibres. Callose is also sometimes found at the surface of the stigma as a response to non-compatible pollen; this may be a form of wound response.

The β1,3-linkage of callose causes it to adopt a helical conformation. This enables it to form gels under some circumstances but it can also form microfibrils in pollen tube walls. It is not always extractable with alkali and may sometimes be found in the α-cellulose fraction.

2.6.7. β1,3-, β1,4-glucans

These unbranched glucans, also known as **mixed-link glucans**, are important cellwall components in cereals and other grasses. The ratio of 1,3 to 1,4 links is between 1 : 2 and 1 : 3, and the usual arrangement of linkages is for single 1,3-linked residues to separate sequences of two, three or four 1,4-linked residues. These glucans are sometimes soluble in water, but often alkali is required for solubilization, perhaps due to covalent links to other wall components. In barley, mixed-link glucan is covalently bound to wall protein, and this may also be true in other grasses.

The β1,4-linked oligosaccharides within the structure tend to form extended, ribbon-like configurations, but the β1,3 links introduce kinks into the structure. The result is an open, irregular conformation, with short linear regions that might be involved in hydrogen-bonding with other β1,4-linked polysaccharides.

2.6.8. Arabinogalactan II

This molecule is found as a component of gymnosperm cell walls, especially in larches. It possesses a highly-branched galactan core, containing β1,3- and β1,6-linked galactose units. Arabinose residues are present on the outer chains, linked by β1,3 bonds. Small amounts of glucuronic acid may also be present.

Similar polysaccharides may be present in a wider variety of tissues, since many suspension-cultured angiosperm cells have been found to secrete arabinogalactan proteins in which the polysaccharide portion is very similar to arabinogalactan II.

2.7 Proteins and glycoproteins

Cell walls contain a variety of different proteins, most of which are glycosylated. The most abundant ones contain an unusual amino acid, hydroxyproline, which is not generally found in the proteins of the protoplast. Such proteins are known as HRGPs (hydroxyproline-rich glycoproteins). The presence of this amino acid makes it relatively easy to pinpoint the cell wall as being the location of such glycoproteins. However, other proteins are present which do not contain hydroxyproline in large amounts. The full range of such proteins is not known, because it is quite difficult to prove that a given protein is located in the cell wall *in vivo*, as opposed to being adsorbed on to the cell wall after homogenization of the tissue.

The most extensively studied family of cell wall glycoproteins is known as **extensin**. These glycoproteins contain a high proportion (around 40%) of their amino acids as hydroxyproline, together with large amounts of serine and lysine. The hydroxyproline frequently occurs as part of a sequence Ser–$(Hyp)_4$, and the hydroxyproline residues are attachment points for tri- and tetra-arabinose oligosaccharides (Figure 2.18). The serine residues are attachment points for single galactose residues. The Ser–$(Hyp)_4$ repeats are separated by an average of about six amino acids, depending on the species. The molecule contains tyrosine residues, which are able to cross-link in the wall to form intramolecular and perhaps also intermolecular covalent bridges. These tyrosine residues are often present as Tyr–Lys–Tyr sequences. About 40–50% of the weight of extensin is protein, and the molecular weight of the uncross-linked monomer is around 40 000. The molecule has a helical secondary structure (the polyproline II structure) and appears as a stiff rod in electron micrographs.

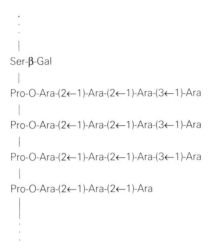

Figure 2.18   The hydroxyproline-rich region of extensin illustrating the tri- and tetra-arabinose oligosaccharides. These side-chains are mixed in linkage, the first three arabinose residues joined by $\beta(1–2)$ bonds, and the outer (fourth) residue linked by a $\beta(1–3)$ bond.

Much of the information on extensin structure is based on gene sequences, identified as extensins on the basis of the characteristic amino acid repeats. Direct study of actual extensin molecules is often difficult since they are hard to extract from the wall, presumably due to extensive cross-linking.

Monocots contain glycoproteins which have some homology with dicot extensins. These glycoproteins include HRGPs which are also rich in threonine (THRGPs) and others which are also rich in histidine and alanine (HHRGPs).

A related protein occurs as part of arabinogalactan proteins (AGPs), the polysaccharide portion of which is similar to arabinogalactan II (section 2.6.8). Arabinogalactan proteins are found in the extracellular fluid of many suspension-cultured tissues and are thought to be components both of the cell wall and of extracellular secretions in the multicellular plant. The protein portion, which normally comprises 2–10 % of the weight, is rich in hydroxyproline, serine, alanine and glycine. The arabinogalactan is probably linked to hydroxyproline through a β-galactose linkage.

Another related group of proteins contains a high proportion of proline as well as, usually, of hydroxyproline (the proline-rich proteins, or PRPs). Some of these proteins are found in normal plant cell walls, while others are **nodulins**, i.e. they are specifically found in the cell walls of nitrogen-fixing root nodules.

The glycine-rich proteins (GRPs) form a further group of wall-associated proteins, probably located at the wall–plasma membrane interface. These have repeated Gly–X motifs, the X often also being Gly. Glycine may make up two-thirds of the amino acids. These proteins are associated with walls at the onset of lignification and may act as nucleation sites for lignin formation.

The enzymes that are located in the cell wall include peroxidase, invertase, cellulase, acid phosphatase, pectinase, pectin methylesterase and malate dehydrogenase. A number of exoglucosidases have been reported, including β-glucosidase, β-xylosidase, α-galactosidase, and β-galactosidase. Endo-β1,4-glucanase is present, and may be involved in cell wall turnover, especially of xyloglucan. Endo-β1,3-glucanase is also present. Much attention has been given to xyloglucan endotransglycosylase (XET), which is found in a wide range of walls and may be involved in insertion of xyloglucan into the wall (section 4.6) and possibly in control of wall extensibility (section 5.5.2).

The cell wall also contains **lectins** (proteins that bind specifically to certain sugars without acting enzymically on them). Some of these lectins, such as the potato lectin, contain domains with amino acid sequences homologous to extensins.

2.8 Lignin and other phenolic compounds

Certain differentiated cell types contain lignin, a phenolic polymer that is laid down after cell elongation has ceased. The precursors of lignin are the three aromatic alcohols: coumaryl, coniferyl and sinapyl alcohols (Figure 2.19) which give rise to the p-coumaryl, guiacyl and sinapyl propane subunits, respectively. These precursors are linked by a wide variety of bonds in the final polymer. The linkages include carbon–carbon bonds, in which one or both carbons may be aromatic. The linkage pattern is irregular, due to the non-enzymic nature of the polymerization process (Chapter 4). A small part of a representative lignin molecule is shown in Figure 2.20. Polymerization can continue to occur as long as activated precursors and space in the wall are available, so that the molecule tends to fill all the space in the wall not occupied by macromolecules, displacing water as it does so. The result is a very strong, hydrophobic meshwork which surrounds the other wall components and cements them in place. The overall structure is incapable of plastic extension, and hence growth ceases. Lignin is also an effective barrier to the penetration of both nutrients and pathogens, so that fully lignified cells are dead and provide good protection for the rest of the plant against infection.

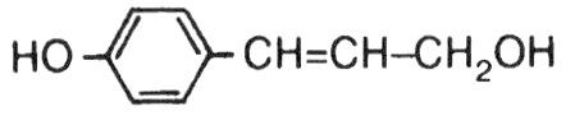

HO— —CH=CH–CH$_2$OH

p–coumaryl alcohol

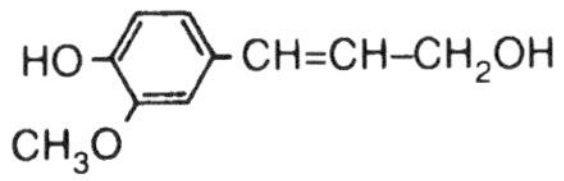

HO— —CH=CH–CH$_2$OH
CH$_3$O

Coniferyl alcohol

CH$_3$O
HO— —CH=CH–CH$_2$OH
CH$_3$O

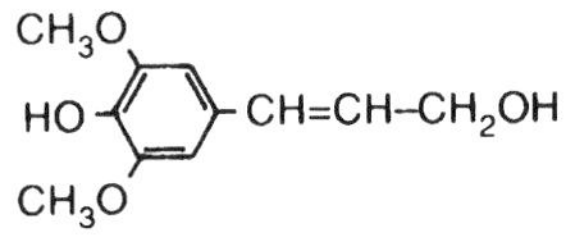

Sinapyl alcohol

Figure 2.19   The aromatic alcohols that are precursors in the synthesis of lignin

Lignin is found principally in sclerenchyma and in the tracheids and vessels of the xylem. It is also found in other cells in response to infection or certain other external stimuli. The chemical constitution of lignin varies to some extent in different taxa: gymnosperms contain a high proportion of guiacyl subunits, while dicotyledonous angiosperms have about equal amounts of guiacyl and syringyl propane subunits. Lignin from monocots contains roughly equal amounts of all three subunits.

In addition to lignin, other phenolic compounds may be present. The most important of these is ferulic acid (Figure 2.21), which is esterified to arabinose and galactose in pectins and may have an important role in the cross-linking of pectins. In cereals, which have only low amounts of pectin, ferulic acid appears to be linked to the arabinose of arabinoxylans. Also found in cereal primary walls is *p*-coumaric acid; both ferulic acid and *p*-coumaric acid are associated with lignin in cereal secondary walls. These phenolic acids are sometimes found as the cyclic dimers, truxillic and truxinic acids, in which four-membered (cyclobutane) rings have formed by cycloaddition.

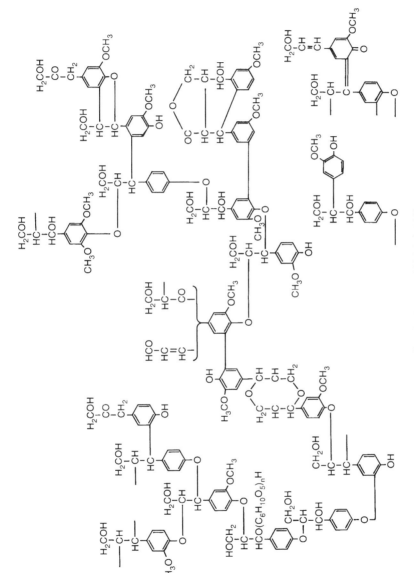

Figure 2.20 Partial structure of spruce lignin as depicted by Freudenberg and Neish (1968)

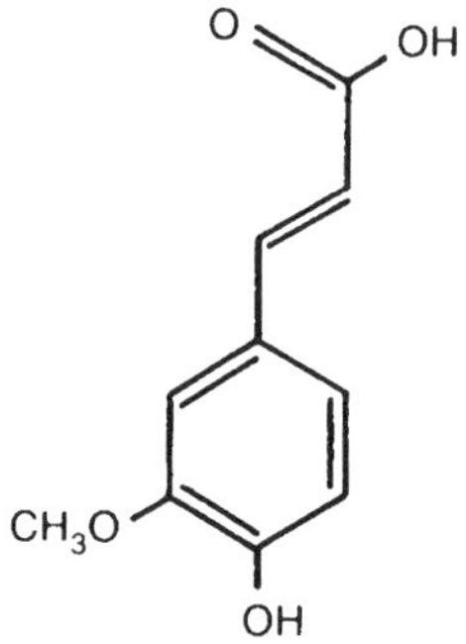

Figure 2.21   Ferulic acid

Summary

The cell wall is laid down as a series of layers. Each consists of a microfibrillar phase (cellulose) and a matrix phase. The matrix phase contains a variety of polysaccharides, which may be divided into pectins and hemicelluloses. The pectins include rhamnogalacturonans, galacturonans and polymers of galactose and arabinose. Hemicelluloses include xylans, glucomannans, mannans, xyloglucans and non-cellulosic glucans. The wall matrix also contains proteins, glycoproteins and phenolic compounds, including lignin.

References

Bacic, A., Harris, P.J. and Stone, B.A. (1988) Structure and function of plant cell walls, in *The Biochemistry of Plants*, Vol. 14, (ed. J. Preiss), Academic Press, New York, pp. 297–371.

Freudenberg, K. and Neish, A.C. (1968) *Constitution and Biosynthesis of Lignin*, Springer, Berlin.

Gidley, M.J. (1992) High-resolution solid-state NMR of food materials. *Trends in Food Sci. & Tech.*, **3**, 231–236.

Levy, S., York, W.S., Stuike-Prill, R. *et al.* (1991) Simulations of static and dynamic molecular configurations of xyloglucan – the role of the fucosylated side-chain in sugar-specific side-chain folding. *Plant J.*, **1**, 195–215.

McCann, M.C., Hammouri, M., Wilson, R. *et al.* (1992) FTIR microspectroscopy is a new way to look at plant cell walls. *Plant Physiol.*, **100**, 1940–1947.

Meyer, K.H. and Misch, L. (1936) Positions des atomes dans le nouveau modele spatial de la cellulose. *Helv. Chim. Acta*, **20**, 232–244.

Nishitani, K. and Nevins, D.J. (1991) Glucuronoxylan xylanohydrolase. A unique xylanase with the requirement for appendant glucuronosyl units. *J. Cell Biol.*, **266**, 6539–6543.

Preston, R.D. (1974) Plant cell walls, in *Dynamic Aspects of Plant Ultrastructure*, (ed. A.W. Robards), McGraw-Hill, Maidenhead, pp. 256–309.

Selvendran, R.R. (1985) Developments in the chemistry and biochemistry of pectic and hemicellulosic polymers. *J. Cell. Sci.* (Supplement 2), 51–88.

Further reading

Carpita, N.C. and Gibeaut, D.M. (1993) Structural models of primary cell walls: consistency of molecular structure with the physical properties of the walls during growth. *Plant J.*, **3**, 1–30.

Fincher, G.B. and Stone, B.A. (1983) Arabinogalactan proteins: structure, biosynthesis and function. *Ann. Rev. Plant Physiol.*, **34**, 47–70.

Keller, B. (1993) Structural cell wall proteins. *Plant Physiol.*, **101**, 1127–1130.

Kieliszewski, M.J. and Lamport, D.T.A. (1994) Extensin: repetitive motifs, functional sites, post-translational codes and phylogeny. *Plant J.*, **5**, 157–172.

McNeil, M., Darvill, A.G., Fry, S.C. and Albersheim, P. (1984) Structure and function of the primary cell walls of plants. *Ann. Rev. Biochem.*, **53**, 652–683.

Marchessault, R.G. and Sundararajan, P.R. (1983) Cellulose, in *The Polysaccharides*, Vol. 2 (ed. G.O. Aspinall), Academic Press, New York, pp. 11–95.

Showalter, A.M. (1993) Structure and function of plant cell wall proteins. *Plant Cell*, **5**, 9–23.

3 Cell wall architecture and the skeletal role of the cell wall

3.1 Wall types

The previous chapter describes the great variety of different molecules that occur in the cell wall. Most of the structural details have been discovered by studies on extracted, partially purified components. One of the most exciting challenges of current cell wall research is to understand the architecture of the intact wall, i.e. how all the components interact to form a living, dynamic structure.

To do this, a wide range of experimental approaches are now available. Cross-links between wall polymers can be revealed by comparing different kinds of chemical and enzymic extraction methods for individual components. Immunocytochemistry has begun to show up the details of polymer distribution in the wall. Freeze-slamming can preserve structure well enough to permit the relative positions of individual components to be seen by electron microscopy. Infra-red and NMR spectroscopy and atomic force microscopy are providing much information on polymer conformation, orientation and mobility in intact walls.

The results of this work have underlined the fact that wall architecture varies in different types of cell and in the different layers of the wall. As noted earlier (section 2.1), walls may be divided into three major layers: the middle lamella, the primary wall and (when present) the secondary wall. The composition of the primary and secondary walls varies in different species and different cell types, especially with regard to the matrix polysaccharides (see Bacic *et al*., 1988, for a detailed survey).

Two major types of primary wall are known. The primary walls of gymnosperms, dicots and most non-graminaceous monocots have a broadly similar matrix composition, the major polysaccharides being pectin and xyloglucan. These are sometimes known as Type 1 primary walls. The Gramineae (cereals and grasses) and a few other monocot families have primary walls with little pectin and with arabinoxylans and β1,3,β1,4 glucans as the major hemicelluloses (Type 2 primary walls). Differences that can be found in the ratios of matrix and other cell wall components are highlighted in Figure 3.1.

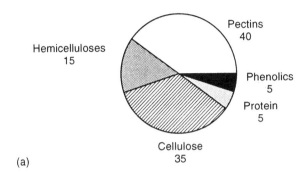

(a)

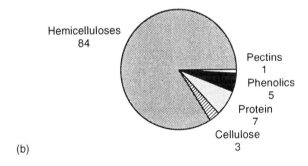

(b)

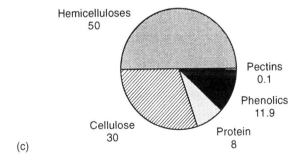

(c)

Figure 3.1 Relative proportions of the components within the cell walls: (a) fruit and vegetable; (b) cereal endosperm; (c) cereal bran.

The matrix is more variable in secondary walls. For lignified secondary walls, a major distinction exists between gymnosperms, which contain glucomannan as the main matrix polysaccharide, and dicots, in which it is replaced by 4-O-methylglucuronoxylan. This is the chief chemical difference between softwoods and hardwoods. Less is known about the secondary walls of monocots; once again, the Gramineae seem to differ from most other monocots, which resemble dicots.

In addition to these broad variations, certain specialized cells have very distinctive wall compositions. This applies to cells in which the wall has a major storage function (Chapter 9), to phloem sieve tubes (Chapter 6) and to cells with specialized roles in reproduction (Chapter 8).

3.2 Cross-links between wall polymers

Cell wall components are much easier to study after extraction than *in situ*, and therefore the cross-links between them in the intact wall are not fully understood. However, a number of such cross-links have been identified, some covalent and some non-covalent. It is difficult at present to assess their relative abundance, so the following paragraphs are principally a qualitative, rather than quantitative, description of the cross-linkages.

3.2.1 Covalent linkages

The distinction between covalent linkages between wall polymers and those within them is not clear-cut. However, certain covalent links are placed in the former class on the grounds of their relative infrequency in the wall. Others are thus classified because they are different in kind from the covalent links normally considered to be within the polymers, or because they are formed much later than the intra-polymeric bonds.

The glycosidic linkage between arabinogalactan and the backbone of rhamnogalacturonan I (section 2.5) is something of a borderline case, since it occurs relatively frequently, especially in older walls. However, it can be considered as an interpolymeric bond, since the arabinogalactan portion of RG I appears to be synthesized independently and, in some cases, to become attached to the backbone of RG I as the walls age.

There is some evidence for covalent bonding between xyloglucan and the arabinogalactan of RG I. This evidence comes from studies in which endopolygalacturonase or endoglucanase treatment of suspension-cultured sycamore cell walls produced polysaccharide fragments containing covalently linked portions of xyloglucan, arabinogalactan and rhamnogalacturonan I molecules. The linkage between xyloglucan and arabinogalactan is not known but is likely to involve the reducing end of the xyloglucan backbone. It probably only occurs in less than 10% of the total xyloglucan of the wall.

There is good evidence for covalent cross-linking of pectin via diferulic acid (Figure 3.2). Ferulic acid is esterified to arabinose and galactose in pectin, and two ferulic acid units can be linked by peroxidase activity to form a diphenyl bond. Ether linkages between ferulic acid and pectin may also occur.

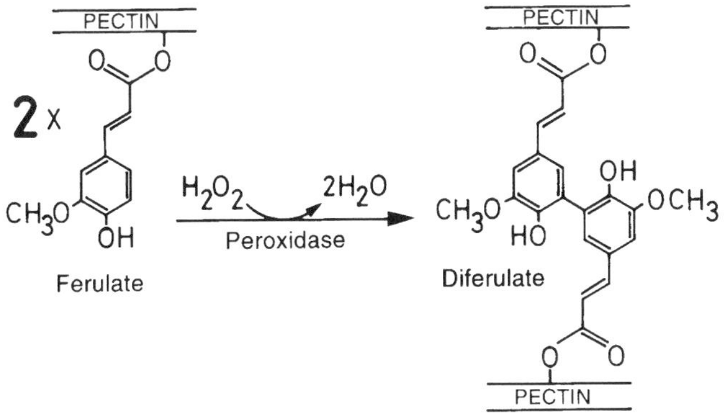

Figure 3.2   Formation of diferulic acid cross-links between pectins.

A similar, peroxidase-catalysed bond may occur between tyrosine residues of extensin. This bond, the phenolic ether linkage of isodityrosine (Figure 3.3), is known to form intramolecularly within extensin, and may also occur intermolecularly. Tyr–Lys cross-links have also been suggested as possible intermolecular bonds (Kieliszewski and Lamport, 1994). Cystine (disulphide) bonds between proteins may also occur, but cysteine is relatively rare in wall proteins.

Ester bonds may occur between polysaccharides (pectin–cellulose esters have been suggested) and between polysaccharide and lignin (where glucuronoxylan–lignin esters are the likeliest possibility).

A number of further cross-links probably remain to be discovered, since covalent associations are found between various polymers extracted from the wall which cannot be assigned to any known cross-link. These include pectin–protein, pectin–xylan, xylan–xyloglucan and proteoglycan–polyphenol complexes, and also more complicated associations such as xylan–pectin–protein, xyloglucan–pectin–protein–polyphenol and xylan–protein–polyphenol.

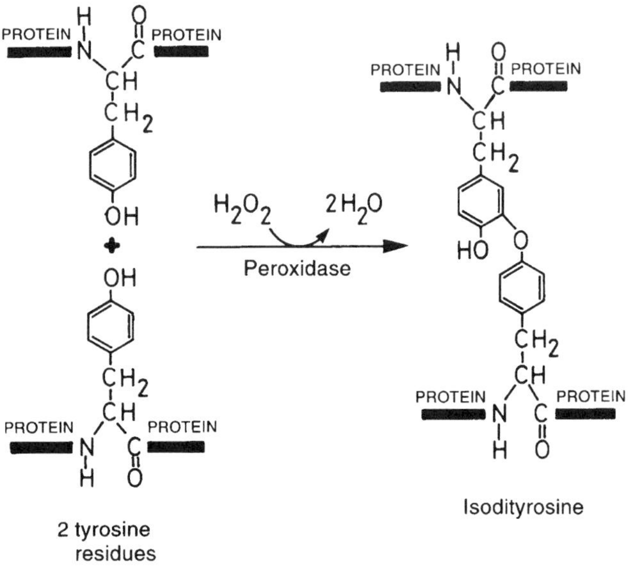

Figure 3.3   Formation of isodityrosine cross-links between cell-wall proteins

3.2.2 Non-covalent linkages

The polyhydroxylic nature of the main wall polymers makes it likely that a great many hydrogen bonds will form in the wall. These become structurally significant when a substantial number form between two macromolecules. We have already described the multiple hydrogen bonding within cellulose microfibrils. The large number of hydrogen bonds which form between the microfibril surface and many hemicelluloses is probably a most important factor in anchoring the microfibrils into the matrix. These hydrogen bonds are wholly or partly responsible for the insolubility of xylans, xyloglucans and glucomannans in the wall, and in the case of xyloglucan and xylan the bonding to cellulose has been demonstrated *in vitro* (Figure 3.4). It is likely that hydrogen bonds also form between hemicelluloses in the matrix. For instance, pairs of xyloglucan molecules (which only have one surface available for hydrogen-bonding) could hydrogen-bond within the matrix. Two or more xylan molecules can form hydrogen-bonded aggregates if the distribution of side-chains allows it. The 'cellulosic' regions of mixed-link glucans (i.e. the contiguous stretches of β1,4-linked glucoses) could also form hydrogen-bonded pairs or aggregates.

All the hydrogen bonds described above are likely to be strengthened in the absence of water, since polysaccharide–water hydrogen bonds

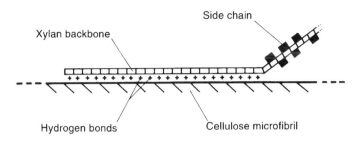

Figure 3.4 Schematic diagram depicting the hydrogen bonding between xylan and cellulose. Note that where the side-chains are attached to the xylan backbone, hydrogen bonding between the xylan and cellulose is prevented.

will compete with polysaccharide–polysaccharide ones. Hence lignified walls, in which the water content is low, will show enhanced strength of hydrogen-bonding compared with non-lignified ones. Ionic bonds are likely to form in the case of those polymers which contain charged groups. There is probably ionic bonding between extensin, with a net positive charge, and the negatively charged galacturonans. The charged groups occur sufficiently frequently in both polymers for some degree of concerted binding to be likely, and both polymers are likely to have regions of ordered secondary structure.

Concerted ionic binding is likely to occur also between two stretches of contiguous galacturonic acid residues, with calcium ions acting as ionic bridges between negatively charged galacturonate residues (Figure 3.5). About 15–20 contiguous galacturonic acid residues are probably needed in each chain for stable complexes to be formed. Such stretches of galacturonic acid residues occur in homogalacturonan and RG I; concerted binding will be interrupted in either case by methyl esterification and, in the case of RG I, by rhamnose residues. Regions of concerted binding, called **'egg-box' linkages**, are likely to be of limited length, but may occur frequently enough to give rise to a gel-like structure in the wall, similar to that found in isolated pectin and in the fruit extracts used in jams and jellies. Other charged groups, such as the glucuronic acid of xylans, may also participate in ionic bonding, but here the bonding is less likely to be concerted.

Hydrophobic forces and hydrogen bonds between methylated galacturonans are thought to be important in the gelling of isolated pectins, and may occur also in the cell wall. Hydrophobic forces may occur between protein molecules and between proteins and lignin.

Since nearly all cell wall macromolecules are large and many tend to take up an irregular, extended conformation, it is likely that spatial entanglement will help to prevent relative movement between them and hence contribute to the strength of the wall. Non-cellulosic glucans

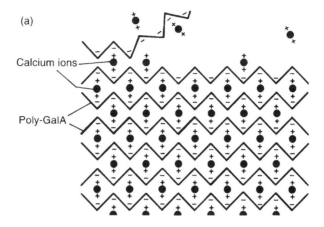

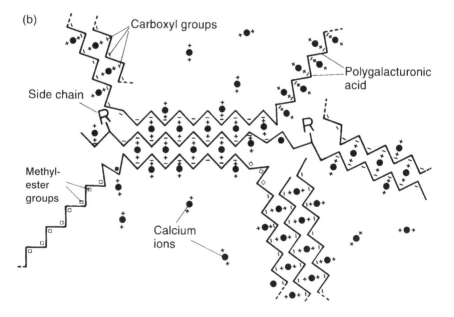

Figure 3.5 Egg-box model of Morris *et al.* (1982) showing maximum cross-linking between unsubstituted polygalacturonic acid (a) and the factors that are likely to influence such cross-linking within the cell wall (b).

tend to form helices or loose coils, due to the kinked β1,3 links in the backbone. The rhamnose units of the pectin backbone introduce kinks into the extended zigzag structure of the polygalacturonic acid backbone. The side-chains in pectin and xyloglucan may bring about frictional resistance to relative movement of polymers in the wall. Such factors are hard to assess, since even our current limited knowledge of wall polysaccharide conformation is principally derived from observations on extracted polysaccharides or from calculations of energy states of isolated polysaccharides. Recent work using FTIR spectroscopy indicates that major changes in conformation occur when pectins are extracted from the wall (McCann *et al.*, 1992). Thus an assessment of the degree of spatial entanglement of wall polymers awaits further studies of polymer conformation *in muro*.

Some of the cell wall proteins are thought to be lectins, i.e. proteins which bind non-covalently to specific sugar residues. For instance, the arabinogalactan-proteins are lectins, binding to β-linked sugars. There is evidence that these AGPs bind to pectin, so the bonding may be to the β-linked galactose residues in pectin side-chains. Once again, it is hard to assess the importance of such binding.

3.3 The cell wall as a set of interlinked networks

The overall structure of the cell wall must be extremely complicated, given the great complexity and variety of the components described above. A number of investigators have attempted to produce models of the whole cell wall. Two early models, those of Keegstra *et al.* (1973) (modified later by Albersheim and co-workers) and of Lamport and Epstein (1983) (Figure 3.6) have been quite influential. Both these models highlight particular intermolecular linkages. The Keegstra model contained covalent links between xyloglucan and RG I and concerted hydrogen bonding between xyloglucan and cellulose; this led to indirect cross-linking of cellulose microfibrils through a series of hydrogen bonds and covalent bonds in the matrix. The Lamport model stressed covalent linkages between protein molecules leading to entanglement of cellulose microfibrils in a lattice of extensin molecules lying at right angles to the microfibrils (a 'warp–weft' arrangement) (Figure 3.6).

To do justice to all the components and their interactions in even one type of cell wall in a single comprehensible model is now probably impossible. With the present degree of knowledge, it is perhaps best to visualize the cell wall as containing a number of polymer networks which, when superimposed upon one another, interact further to give rise to the whole, complex structure. The properties of each network

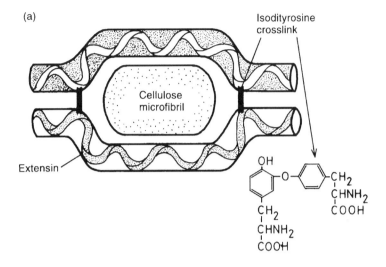

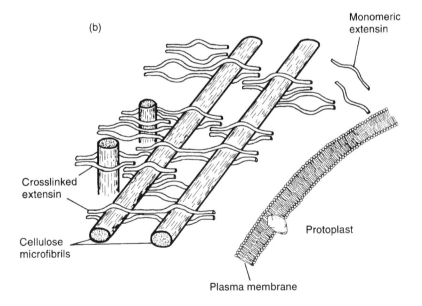

Figure 3.6   Cell wall model of Lamport and Epstein (1983). (a) Proposed mechanism by which extensin might entrap cellulose microfibrils through isodityrosine cross-links. (b) Diagrammatic representation of this concatenation in wall architecture.

are now being explored and the following sections indicate what is known of their properties. The first two networks, composed of polysaccharides, are relatively hydrophilic and may therefore be regarded as gels, holding a significant amount of water between the polymer chains. Such gels typically contain 'junction zones', where complementary configurations permit the polymers to bond to each other through noncovalent bonds. The polymer regions not involved in the junction zones will be in an open, hydrated state. By contrast, the lignin network is extremely hydrophobic, and tends to exclude water from the wall.

Further details of the networks can be found in reviews by Carpita and Gibeaut (1993), McCann and Roberts (1991) and Bacic *et al.* (1988).

3.3.1 The cellulose–hemicellulose network

In most primary walls other than those of the Gramineae (Type 1 walls), xyloglucan is the main hemicellulose, with glucuronoarabinoxylan as a minor component. Both these polymers hydrogen-bond to cellulose and they are thought to coat the surface of the microfibrils with a monolayer of hemicellulose. Xyloglucan is only able to form hydrogen bonds on one side of the molecule, so the coating is limited to one layer. However, cellulose and xyloglucan are present in about equal amounts in the primary wall, and hence only part of the xyloglucan can be bound directly to cellulose. The remainder is thought to span the gaps between microfibrils, forming 'molecular tethers' which hold the microfibrils in place. The evidence for this comes primarily from freeze-slam electron microscopy. McCann *et al.* (1990) were able to visualize these intermicrofibril regions of xyloglucan in a number of cells (Figures 2.2, 3.7). The cross-linking molecules seen in these electron micrographs have been shown by selective extraction methods to be xyloglucans. The length of xyloglucans can be from 20 to 700 nm, with most being around 200 nm long. This is more than enough to span the gap between microfibrils (typically 20–40 nm in dicots) and to bind to sections of microfibrils at each end. Sections of xyloglucan in the intermicrofibril space may bind to each other or to other matrix components, or may interlock with each other as suggested by Carpita and Gibeaut (1993). The degree to which any given section of xyloglucan can bind to microfibrils or other molecules may be governed by the presence or absence of acetyl or arabinose groups (section 2.6.5).

In primary walls of the Gramineae (Type 2 walls), glucuronoarabinoxylans (GAXs) are the major hemicellulose. They may cross-link microfibrils in a similar manner to that proposed for xyloglucans in Type 1 walls. In growing walls, GAXs have a high degree of backbone substitution, which is likely to limit hydrogen-bonding. In more mature walls, the arabinose content decreases, probably due to the action of arabinosidases, and this may increase the binding of GAX to cellulose and thereby decrease wall extensibility. Actively growing Type 2 walls also contain significant amounts of mixed-link glucan. Here again, stretches of contiguous β1,4-linked glucoses may hydrogen-bond to the surface of microfibrils, with the binding limited by the kinks introduced by the β1,3 links. The regions of these molecules in the intermicrofibrillar space may hydrogen-bond to each other, or to the xylans, in the same way. They may also be entangled with each other and with other polymers as a result of their irregular conformation.

3.3.2 The pectin network

The pectic polymers form a network which is largely independent, structurally, of the cellulose–hemicellulose network. Pectins can be removed from the wall without greatly affecting the appearance of the cellulose–hemicellulose network in freeze-slammed specimens; conversely, extracted pectin can form a variety of different types of gel *in vitro*. The bonds holding the pectin network in place in the wall may be ionic, i.e. of the egg-box type (Figure 3.5). Alternatively, when the pectins are heavily methylated or when calcium levels are relatively low, the junction zones may be held together by hydrophobic or hydrogen bonds (Wilkinshaw and Arnott, 1981). In more mature walls, diferulate cross-links between pectins may make the network more rigid, contributing to the overall increase in wall rigidity. As mentioned earlier (section 3.2.1), there may be some covalent links between pectin and other wall components, but they are probably relatively rare in growing walls.

Though the pectin network is distinct from the other networks, it is clearly influenced by them. FTIR studies show that the conformation of pectin in the wall is substantially different from that of extracted pectin. Pectin is sometimes found to be strongly bound to cellulose. Thus other wall components must have some degree of structural interaction with pectin.

The pectin network also interacts functionally with the other networks. Most significantly, pectin appears to control the pore size in the wall, and hence the movement of macromolecules through the wall. Therefore, it may control the access of enzymes to their substrates, and thus make an indirect but important contribution to the control of wall mechanical properties.

Generally the cellulose–hemicellulose network is thought to provide the main structural strength in growing cell walls. However, in exceptional circumstances pectins may assume this role. Suspension-cultured tomato and tobacco cells have been grown in medium containing dichlorobenzonitrile (DCB), a herbicide that specifically inhibits cellulose synthesis. Under these conditions, the cells adjust by growing with a very low level of cellulose in their walls. The cellulose–hemicellulose network is replaced by a tightly cross-linked pectin network, which becomes the main structural component of the wall (Shedletzky *et al.*, 1992).

3.3.3 The extensin network

The hydroxyproline-rich glycoprotein, extensin, forms a third network in the cell wall. Extensin is synthesized and secreted through the plasma membrane as a water-soluble precursor protein. Once in the wall, it is linked into an insoluble extensin network. This network is structurally independent of the polysaccharide networks; this has been shown by treatment of the wall with hydrogen fluoride, which breaks glycosidic bonds but leaves the polypeptide backbone of extensin intact. After such treatment, the extensin network remains insoluble, showing that the cross-links between extensin molecules do not involve wall polysaccharides or the oligosaccharide side-chains of extensin.

The nature of the cross-links is not known for certain. Isodityrosine bonds form between tyrosine residues within an extensin molecule, but it is not certain that they also link different extensin molecules together. Tyr–Lys links are another possible candidate for this cross-linking.

Extensins are elongated molecules, and they are orientated perpendicular to the plasma membrane once they are linked into the insoluble network. This means that they are also perpendicular to the cellulose microfibrils and their associated hemicelluloses, which are laid down parallel to the plasma membrane. They may thus hold the layers together (Figures 3.7 and 3.8). Although this glycoprotein was originally named extensin because a role in catalysing cell growth was proposed for it, it is now thought to have the opposite effect. Generally, high levels of wall extensin are correlated with decreasing or zero cell growth, and extensin may therefore make the wall more rigid (section 5.5.3).

Though structurally independent of polysaccharides, the extensin network may form both non-covalent and covalent bonds with the other wall networks. Ionic bonds may form with pectins (section 3.2.2) and covalent bonds may form with a number of other wall components, though no details are known (section 3.2.1).

3.3.4 The lignin network

The lignin network (Figures 2.20 and 3.9) is the last to be formed and, like the wall networks already described, its basic structure is independent of the other networks. Artificial lignins can be formed by dehydrogenation of lignin precursors and these share many of the molecular properties of lignin. However, there is good evidence for covalent bonds

(a)

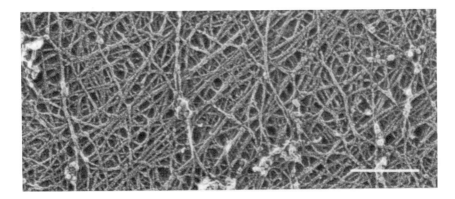

(b)

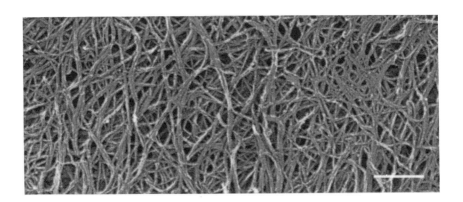

(c)

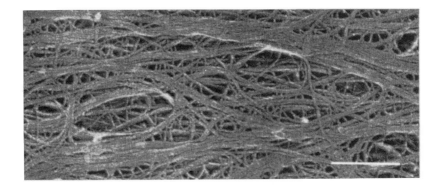

(d)

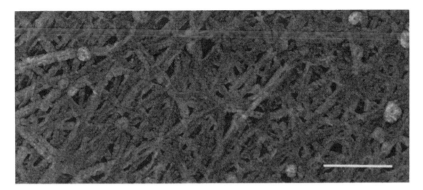

Figure 3.7   Onion cell wall fragments before and after sequential extraction with selected aqueous solvents. Images were obtained by TEM after implementation of a fast-freeze, deep-etch, rotary-shadowed replica technique; they have been printed in reverse contrast. (a) Unextracted cell wall; followed by (b) extraction of pectic polysaccharides in CDTA and Na_2CO_3; followed by (c) extraction in 1 mol dm^{-3} KOH; followed by (d) extraction in 4 mol dm^{-3} KOH and acidified chlorite (note that the cellulose microfibrils have swollen to a diameter of 20 nm). Bars, 200 nm.

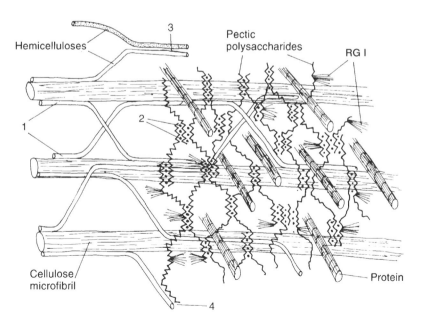

Figure 3.8   Possible interactions between several classes of cell wall polymer and the likely relative orientations.

linking lignin through ether links to ferulic acid, which in turn is ester-linked to polysaccharides. Other covalent bonds to polysaccharide and protein may also occur (section 3.2.1).

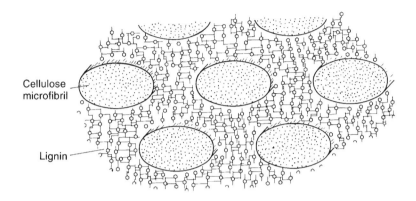

Cellulose microfibril

Lignin

Figure 3.9 Cell-wall model showing the way in which lignin phenolics fill in the spaces between the cellulose microfibrils, thus creating a rigid, impermeable cell wall

Spectroscopic evidence indicates that the lignin structure is quite organized. The planes of the phenyl rings are preferentially orientated in the plane parallel to the plasma membrane. Lignin is deposited between the hemicellulose molecules in lamellae which are also orientated in the same plane.

The presence of the lignin network greatly affects the properties of the other networks. The lignin replaces water, transforming the hydrophilic, hydrated gel into a hydrophobic environment. This increases the strength of hydrogen bonds between polysaccharides, which in turn increases the strength of the cellulose–hemicellulose network. The latter is further strengthened by the low degree of backbone substitution found in the hemicelluloses of secondary walls. In addition, the absence of water will prevent the penetration of wall-loosening enzymes into the wall; it also means that no aqueous 'lubricant' is available to facilitate the 'creep' of wall polymers past one another that occurs during plastic wall elongation (section 5.2.1).

This combination of effects, together with the strength of the lignin network itself, brings about a great increase in the rigidity of the wall. Lignification is generally accompanied by considerable wall thickening and the end result is a wall that has been transformed from a dynamic, extensible structure under close metabolic control into a rigid structure, specialized to provide structural strength (sections 3.5.1 and 3.6).

3.4 Distribution of polymers within the wall

A general picture of polymer distribution within the wall was obtained using histochemistry and the chemical analysis of different types of wall. These approaches showed, for instance, that pectin levels were high in the middle lamella and very low in the secondary wall. They also identified the broad differences in composition between primary and secondary walls (section 3.1).

In recent years, antibodies against wall components have become available (Knox, 1992). These permit localization of wall components much more precisely, by immunocytochemistry. For instance, antibodies against RG I show it to be confined mostly to the middle lamella in suspension-cultured sycamore cells, while antibodies against xyloglucan show that it is located in both middle lamella and primary wall. Monoclonal antibodies show that pectins with different degrees of methylation have different distributions within the wall (Figure 3.10). Monoclonal antibodies against calcium-linked 'egg-box' dimers of polygalacturonic acid indicate that pectin junction zones with this structure are largely confined to the expanded middle lamella regions found at cell corners. Antisera against callose and arabinogalactan 1 have shown that both are found in the early cell plate at cell division. Antibodies against arabinogalactan-proteins (AGPs) suggest that they are located at the plasma membrane–wall interface, and may therefore be involved in anchoring the wall to the plasma membrane. Similarly, antibodies against glycine-rich proteins show that they have the same location.

Antibodies can also reveal subtle differences in wall structure between otherwise similar cells. AGPs at the surface of root apical meristem cells in carrot are developmentally regulated, reflecting the position of the cells in the meristem and hence anticipating developmental differences that will occur amongst the daughter cells derived from these meristematic cells (Smallwood *et al.*, 1994).

3.5 Chemical and physical properties of the whole cell wall

3.5.1 Mechanical strength

The mechanical strength of non-lignified wall is principally due to the strength of the cellulose microfibrils. In lignified walls, the lignin also contributes considerably to the strength. In both types of wall, the strength is greatest in the direction parallel to the microfibrils, and least in the direction perpendicular to them. Hence the orientation of the microfibrils is extremely important in determining the behaviour of cell

(a)

(b)

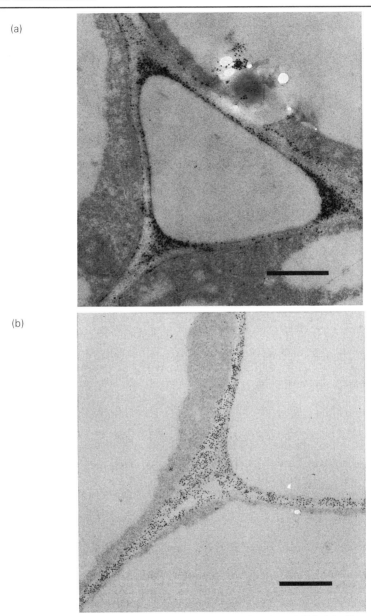

Figure 3.10 Immunogold electron micrographs of JIM5 and JIM7 labelling cell walls of cortical cells of the carrot root. In these micrographs, monoclonal antibodies recognizing unesterified (JIM5 antibodies) and methyl-esterified (JIM7 antibodies) epitopes of pectin were used to locate these epitopes in the cell wall of carrot root by indirect immunogold electron microscopy, and showed differential location of the epitopes within the cell walls. (a) Gold particles labelled by JIM5 (black dots) were located to the cell wall adjacent to the plasma membrane, middle lamella, and the surface of cell walls exposed at the intercellular spaces. (b) Gold particles labelled by JIM7 were distributed evenly throughout the cell wall. Bars, 500 nm.

walls under stress. It is a common feature of cell walls that the orientation of the microfibrils is different in different wall layers, with the overall effect that the structure is strong no matter from what direction the stress comes. Thus the wall is constructed rather in the same way as, at the macroscopic level, plywood is constructed, the direction of the grain in the plywood layer being analogous to the direction of the microfibril orientation in the wall layers (Figure 3.11).

This principle of variable orientation is seen most clearly in secondary walls. In tracheids, for instance, the secondary wall is composed of three

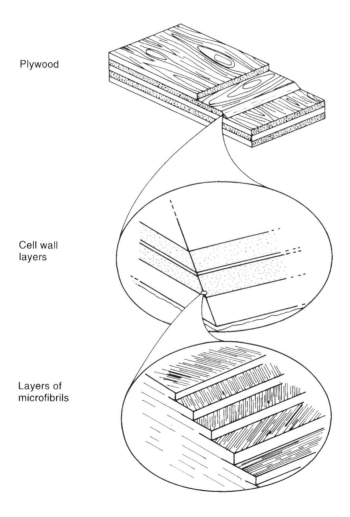

Plywood

Cell wall
layers

Layers of
microfibrils

Figure 3.11 Comparison between the architecture of the cell wall and the structure of plywood. Note the way in which the layers of microfibrils have differing orientations so that, as in plywood, the final structure has multidirectional support.

major layers, with different microfibril orientation in each layer (section 3.6).

In primary walls, there is probably less clear-cut orientation in different layers. There is a tendency for microfibrils to be laid down with a net transverse orientation at the inner face of the wall; this layer is then thought to be passively reorientated to a more longitudinal orientation as the cell extends (Chapter 5). Some primary walls, however, have been shown to have a finitely ordered helicoidal structure over at least part of their thickness. A helicoidal structure consists of a series of layers in which the orientation of the microfibrils changes by a constant and usually small angle from one layer to the next (Figure 3.12). Overall, this type of structure gives uniform mechanical properties in all directions within the plant parallel to the layers. However, it will also be subject to passive reorientation as the cell wall extends during cell growth, giving rise to a net longitudinal orientation.

The behaviour of cell walls under tensile stress is a complicated mixture of plastic and elastic deformation. In the case of primary walls, this behaviour is intimately related to the extension behaviour of growing cells and will be considered in Chapter 5. The much stronger secondary walls show only a limited ability to extend; elastic extension is small, with Young's moduli of the order of 10^9 to 10^{11} μNm^{-2}. Some plastic extension can occur, with breaking strains of the order of 2–15%. However, the forces needed to achieve significant extension are seldom generated within the plant, and hence secondary walls are effectively inextensible.

3.5.2 Charge

The cell wall is normally found to have a net negative charge on its stationary phase, owing to a preponderance of uronic acid residues. These uronic acids are chiefly the galacturonic acid of pectins in primary walls and the glucuronic acid of xylans in secondary walls. The negative charges may be partially balanced by positively charged proteins. The remaining counter-ions are small, predominantly calcium. Hydrogen ions may also be present in significant quantities, the pH of the wall being generally between 6 and 4.

3.5.3 Porosity

In hydrophilic, non-lignified cell walls, the mean separation of the relatively immobile polymers is such that small molecules can penetrate readily, provided they are not immobilized by electrostatic or other forms of binding to the polymers. Thus sucrose, growth substances and most amino acids penetrate without difficulty. Some movement of small

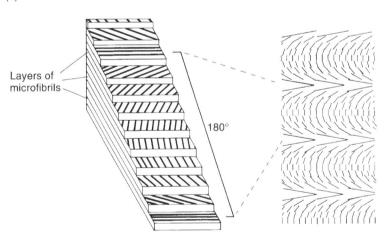

Figure 3.12 Helicoidal patterns within the cell wall. (a) TEM of a section through cell walls of *Echinochloa colonum* (L.) collenchyma illustrating the occurrence of helicoidal patterning. In the top cell wall the section has cut obliquely through the layers of microfibrils, therefore resulting in the helicoidal pattern (see (b) for reasoning). In the lower cell wall, the section is in the same plane as the microfibrils and does not show this effect. Bar, 1 mm. (b) Diagram showing how helicoidal patterns arise from different layers of microfibrils.

proteins and polysaccharides is also possible, the diameter of pores in the wall being generally in the range 3.5–5.5 nm. Large proteins are generally immobile; the borderline for proteins is between 10 000 and 50 000 daltons. However, there are some examples of larger proteins penetrating the wall, and some cells are specialized to secrete large proteoglycans (e.g. root-cap cells). Hence certain cells may possess a number of broader channels through the cell wall, to permit the secretion of macromolecules.

3.5.4 Hydrophilic/hydrophobic character

Primary walls are hydrophilic, with continuous channels of water extending through them, permitting movement of small molecules (see above). However, lignification replaces the water content by an extremely hydrophobic network of phenylpropanoid subunits, converting the wall to a hydrophobic state. On the epidermis and endodermis, other hydrophobic encrustations convert the wall into a hydrophobic structure (Chapter 6).

3.6 Mechanical strength as a specialized property of certain cells

The cell wall plays a role in providing mechanical strength in all walled plant cells. In those cells that possess only an unthickened primary wall, the strength depends also on the turgor of the cell. The wall itself only gives limited support, but when it is tautened by the outward pressure of a turgid cytoplasm the strength of the system greatly increases. Hence the decrease in turgor which occurs under conditions of water stress often results in wilting.

Parenchyma cells which have unthickened primary walls may nevertheless show some specialization in relation to imposed stress. The vertically orientated walls of stem parenchyma show a tendency to transverse orientation of microfibrils, which permits the cell to bend without breaking. In roots, on the other hand, the microfibrils have a more steeply pitched, helical orientation, which gives greater resistance to extension forces that occur when the root is pulled.

In many cells the wall is further thickened and strengthened, to provide a structure which can act independently of turgor pressure in providing structural support. This permits the plant to withstand the major mechanical stresses imposed by gravity and by environmental influences such as wind and snow. Furthermore, tall-growing plants must be able to withstand the hydraulic stresses involved in transporting water to the highest points. The following sections summarize the wall structure in some important strength-providing cell types.

3.6.1 Collenchyma

Collenchyma is a supporting tissue found towards the periphery of herbaceous stems and petioles, and in the ribs of leaves. Often bundles of collenchyma form distinctive ribs or ridges at the outer edges of stems and petioles – well-known examples are the edible petioles of celery and rhubarb.

Collenchyma cells have thickened primary walls, containing pectin, hemicellulose, protein and cellulose but no lignin. The thickening is unevenly distributed, being greatest either at the corners or along the tangential walls (Figure 3.13). Wall thickenings are deposited while the cell is still growing, so that the wall increases in area and thickness simultaneously. The deposition of wall thickenings occurs in part as a response to stress, since it can be stimulated by application of mechanical stress to growing plant parts. The wall often shows clear-cut layering, with alternation of layers having a predominantly transverse microfibril orientation with those having a predominantly longitudinal orientation.

Collenchyma cells retain their cytoplasm even when mature, in contrast to other supportive wall types. Thus they are able to respond to the changing needs of a developing plant organ by laying down further wall layers if needed.

3.6.2 Sclerenchyma

Sclerenchyma cells are thick-walled, lignified cells containing a secondary wall laid down over the primary wall after wall extension has ceased. Both primary and secondary parts of the wall are lignified at maturity. Sclerenchyma cells are usually divided into two types: **sclereids**, which are approximately isodiametric, and **fibres**, which are greatly elongated (Figures 3.14, 3.15). Intermediate types occur and may take up branched shapes.

In sclerenchyma, cell wall thickening is taken to extremes, with the wall generally occupying at least half the cell volume at maturity. The secondary wall is usually deposited evenly over the whole surface of the primary wall, often leaving only a narrow canal at the centre of fibres. The cells are normally dead at maturity, especially when the walls are very thick.

Fibres are found as supporting cells in and around the vascular tissue, often occurring as multicellular strands. In contrast to collenchyma cells, they are present when elongation of the plant organ has ceased. Within a strand, fibre cells often overlap, which gives increased strength

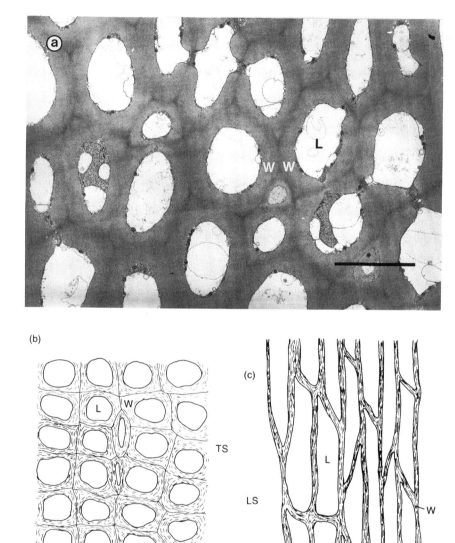

Figure 3.13   Collenchyma cells. (a) TEM of collenchyma cells in TS from *Echinochloa colonum* (L.). Bar, 10 μm. (b) Line drawing of transverse section. (c) Line drawing of longitudinal section. L, lumen; W, thickened primary cell wall.

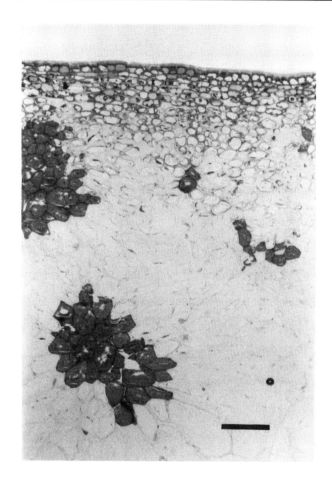

Figure 3.14 Light micrograph showing the presence of sclereid clusters within the parenchyma cortex of Spanish pear (*Pyrus communis* cv Blanquilla). The sclereids immediately beneath the epidermis, together with the resistance of these tissues to ripening (as indicated by maintenance of staining density), provide a relatively tough, protective skin to the pear. Bar, 100 μm.

to the strand as a whole. Some plants, such as flax and jute, contain fibre strands that are commercially important.

Sclereids may also occur as strands or layers, in which case they too have a structural role. This occurs, for instance, in many hard seed coats. In other cases, they occur as small groups of cells, for instance in pear fruit, where they impart a 'grittiness' to the texture (Figure 3.14).

3.6.3 Xylem tracheary elements

Xylem tracheary elements are the water-conducting cells of vascular plants. They are divided into two types: the **tracheids** and the **vessel elements** or vessel members. The major difference between the two types is that water movement between tracheids occurs through **pit-pairs**, while vessel elements have perforations in their end-walls which allow free movement of water from one to another (Chapter 6). Vessel elements may also have perforations and/or pit-pairs in their side walls. In addition to the difference in water movement between the two types of tracheary elements, vessel elements are often wider than tracheids. Tracheary elements are dead at maturity, and contain thick, lignified secondary walls. Usually, the secondary wall is laid down unevenly on the wall surface, giving rise to characteristic patterns of wall thickening (Figure 3.16). Some of these patterns permit some continued cell extension, even though the lignified parts of the wall cannot extend. Examples of this are annular and helical tracheids, in which cell extension can be achieved by extension of the unthickened primary wall which lies between the wall thickenings. In the case of helical wall thickenings, extension leads to a steepening of the helix, as in an extended

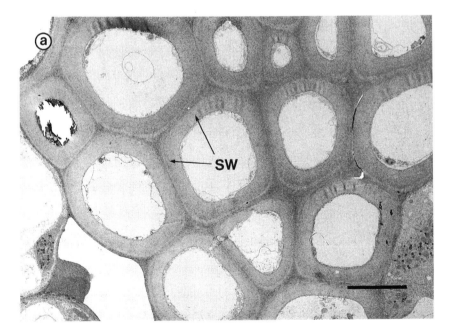

Figure 3.15 Sclerenchyma cells. (a) TEM of a TS of sclerenchyma cells from wheat (*Triticum aestivum*) leaf. Bar, 5 μm. Note the heavily-thickened secondary walls, SW. (b) Line drawing of fibres in transverse (TS) and longitudinal (LS) section; the distance between the arrow heads may be several millimetres. (c) Line drawing of a sclereid (stone cell).

(b)

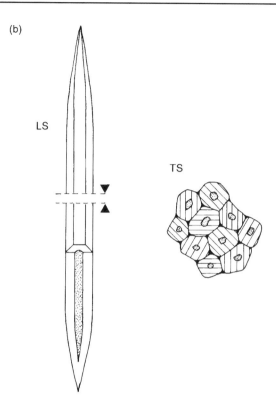

(c)

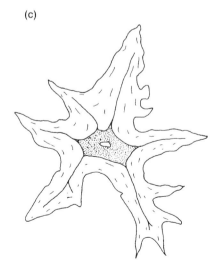

spring. All the other main types of thickening (scalariform, reticulate and pitted) lead to a cessation of elongation. Thus annular and helical thickenings are characteristic of tracheids which form in the youngest, elongating parts of stems and roots and in expanding leaves.

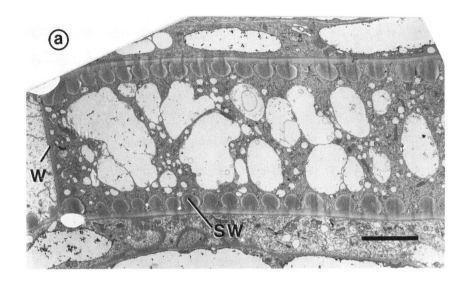

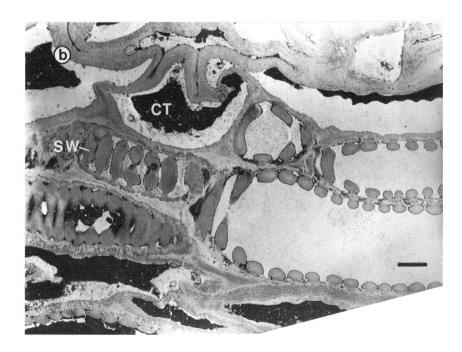

(c)

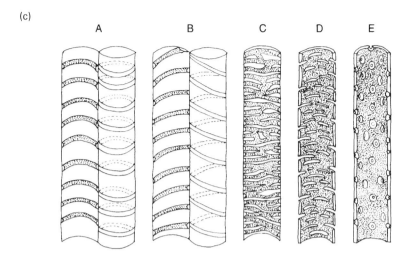

Figure 3.16   Secondary thickening in xylem elements. (a) TEM of an LS through an immature xylem element from *Echinochloa colonum* (L.) showing the recently-formed secondary wall thickenings (SW). Note that the cell is full of cytoplasm and that the end walls (W) have not yet started to break down. Bar, 5μm. (b) TEM of an LS through mature xylem elements of *Camellia sinensis* (L.). The secondary thickening (SW) is evidently annular in form as shown by the elements on the left of the picture. The black shrivelled appearance of the cytoplasm (CT) of the surrounding cells is due to the treatment that the material received prior to fixation. The specimen is, in fact, a dried tea leaf. Bar, 2μm. (c) The types of secondary wall thickening commonly found in xylem elements. A, annular thickening; B, helical thickening; C, scalariform thickening; D, reticulate thickening; E, pitted thickening. Annular and helical thickening allow further cell extension through growth of the primary wall. The other forms of thickening prevent further extension.

The secondary walls of tracheids contain, typically, three distinct layers (Figure 3.17). The microfibril orientation is relatively uniform in each layer but differs from layer to layer. In S1, the outermost layer (next to the primary wall), it is usually a slow helix (i.e. relatively transverse), while in the S2 layer it is relatively longitudinal. The S3 layer then reverts to a more transverse orientation. The interfaces between the three layers have a series of thin (one microfibril thick) lamellae in which the orientation changes by a small angle in each successive layer, so that the transition is achieved by a portion of a helicoid (section 3.5; Figure 3.18).

Some xylem tissues have a special adaptation to resist the force of gravity. These tissues form '**reaction wood**', which is found in branches and in stems which are growing at an angle to the vertical. In conifers, the reaction wood is found on the lower side of the branch or stem, and apparently resists compression. In dicotyledons, the reaction wood is on

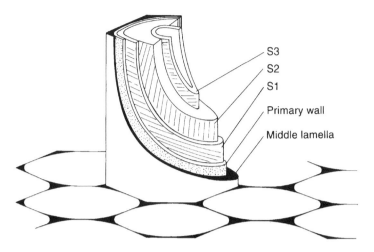

Figure 3.17   S1, S2 and S3 layers of the tracheid cell wall showing the relative orientation of the cellulose microfibrils in each layer.

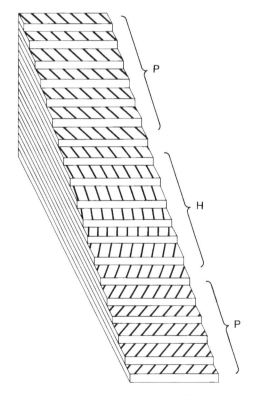

Figure 3.18   Schematic diagram showing the helicoidal (H) transition between layers of parallel (P) cellulose microfibrils.

the upper side and resists tension. In each case the reaction wood differs in structure from ordinary wood; at the cell wall level, conifer reaction wood shows an altered structure in that the S3 layer is usually absent.

Summary

A number of different wall types have been recognized. In each case the components of the matrix are, to some extent, cross-linked by covalent and non-covalent bonds, making up complicated macromolecular networks, which hold the microfibrils in place. Four distinct networks may be present: the cellulose–hemicellulose, pectin, extensin and lignin networks. The microfibrils are orientated in different directions in different wall layers, causing non-uniform mechanical properties along different axes of the cell. The charge, porosity and hydrophobic/hydrophilic character of the wall depend on the composition of the matrix. The walls of certain cell types are specialized for structural strength chiefly by thickening and lignification.

References

Bacic, A., Harris, P.J. and Stone, B.A. (1988) Structure and function of plant cell walls, in *The Biochemistry of Plants*, Vol. 14, (ed. J. Preiss), Academic Press, New York, pp. 297–371.

Carpita, N.G. and Gibeaut, D.M. (1993) Structural models of primary cell walls in flowering plants: consistency of molecular structure with the physical properties of the walls during growth. *Plant J.*, **3**, 1–30.

Keegstra, K., Talmadge, K.W., Bauer, W.D. and Albersheim, P. (1973) The structure of plant cell walls III. A model of the walls of suspension-cultured sycamore cells based on interconnections of the macromolecular components. *Plant Physiol.*, **51**, 188–196.

Kieliszewski, M.J. and Lamport, D.T.A. (1994) Extensin: repetitive motifs, functional sites, post-translational codes and phylogeny. *Plant J.*, **5**, 157–172.

Knox, J.P. (1992) Molecular probes for the plant cell surface. *Protoplasma*, **167**, 1–9.

Lamport, D. and Epstein, L. (1983) A new model for the primary cell wall: concatenated extensin–cellulose network. *Curr. Topics Biochem. Physiol.*, **2**, 73–87.

McCann, M.C. and Roberts, K. (1991) Architecture of the primary cell wall, in *The Cytoskeletal Basis of Plant Growth and Form*, (ed. C. Lloyd), Academic Press, New York, pp. 109–129.

McCann, M.C., Hammouri, M., Wilson, R. *et al.* (1992) FTIR microspectroscopy is a new way to look at plant cell walls. *Plant Physiol.*, **100**, 1940–1947.

McCann, M.C., Wells, B. and Roberts, K. (1990) Direct visualisation of cross-links in the primary cell wall. *J. Cell Sci.*, **96**, 323–334.

Morris, E.R., Powell, D.A., Gidley, M.J. and Rees, D.A. (1982) Confirmations and interactions of pectins. *J. Mol. Biol.*, **155**, 507–516.

Shedletzky, E., Shmuel, M., Tainin, T. *et al.* (1992) Cell wall structure of cells adapted to growth on the cellulose inhibitor DCB – a comparison between two dicots and a graminaceous monocot. *Plant Physiol.*, **100**, 120–130.

Smallwood, M., Beven, A., Donovan, N. *et al.* (1994) Localisation of cell wall proteins in relation to the developmental anatomy of the carrot root apex. *Plant J.*, **5**, 237–246.

Wilkinshaw, M.D. and Arnott, S. (1981) Conformation and interactions of pectins. II. Models for junction zones in pectinic acid and calcium pectate gels. *J. Mol. Biol.*, **153**, 1075–1085.

Further reading

Esau, K. (1977) *Anatomy of Seed Plants*, 2nd edn., John Wiley and Sons, New York.

Fry, S.C. (1986) Cross-linking of matrix polymers in the growing cell walls of angiosperms. *Ann. Rev. Plant Physiol.*, **37**, 165–186.

Jarvis, M.C. (1984) Structure and properties of pectin gels in plant cell walls. *Plant Cell Env.*, **7**, 153–164.

Neville, A.C. (1985) Molecular and mechanical aspects of helicoidal development in plant cell walls. *Bioessays*, **3**, 4–8.

Roberts, K. (1989) The plant extracellular matrix. *Current Opinion in Cell Biology*, **1**, 1020–1027.

Talbott, R.D. and Ray, P.M. (1992) Molecular size and separability features of pea cell wall polysaccharides – implications for models of primary wall structure. *Plant Physiol.*, **98**, 357–368.

4 Cell wall formation

4.1 Stages of wall formation

4.1.1 Cell plate formation

The earliest phase of wall formation occurs as an integral part of cell division. As the two sets of chromosomes separate to form the daughter nuclei during anaphase, small vesicles accumulate and align themselves in the plane of cell division, approximately midway between the two daughter nuclei (Figure 4.1). Starting at the centre of the cell, these vesicles fuse to form the earliest part of the new wall, called the **cell plate**. The cell plate grows from the centre of the cell out towards the pre-existing cell walls, until it eventually fuses with them. Cell division is then complete.

The zone in which the new cell plate forms is called the **phragmoplast**. It is characterized by numerous **microtubules** which lie approximately parallel to one another and perpendicular to the forming cell plate. They are most numerous around the periphery of the growing cell plate and are thought to play a part in guiding vesicles to the edges of the cell plate. Some of the microtubules maintain their position in the phragmoplast while the cell plate forms around them, thus creating channels which preserve cytoplasmic connections between the two daughter cells. These channels are called **plasmodesmata**, and the microtubule may give rise to the **desmotubule** which is found within the plasmodesma (section 6.3).

The vesicles which fuse to form the cell plate contain material which becomes part of the cell plate. The membranes of the vesicles give rise to the new plasma membrane. The chemical composition of the cell plate is not known in detail, since it is difficult to obtain enough of it separated from older wall material for chemical analysis. Cytochemical tests indicate that it contains a high proportion of pectin and perhaps some callose ($\beta(1-3)$glucan). Immunocytochemistry confirms the presence of callose and also of AG1. Microfibrils can be detected, though not close to the growing tip of the cell plate; microfibrils are not seen in the vesicles which give rise to the cell plate.

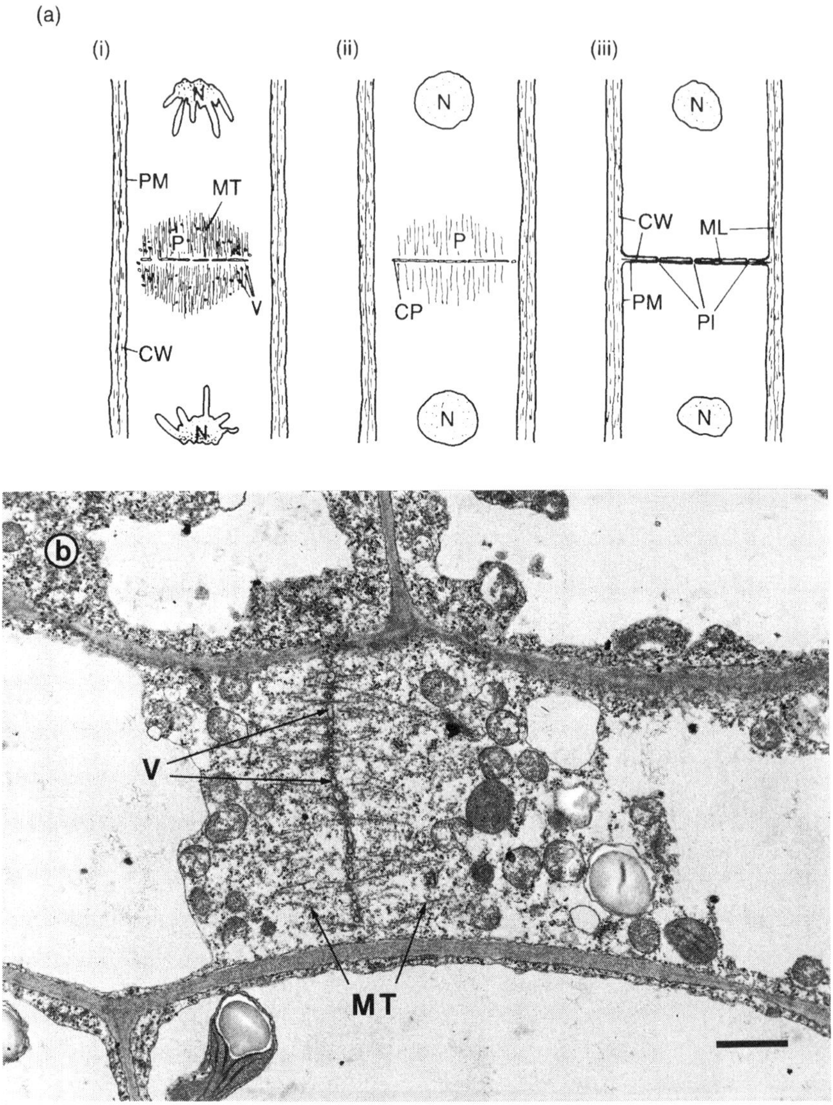

Figure 4.1   The stages in cell wall (CW) synthesis during cell division. (a) Line drawings illustrating cell plate (CP) formation from vesicles of the endomembrane system. In (i), the nuclei (N) have separated and the phragmoplast (P), which consists of many microtubules (MT), is engaged in arranging vesicles (V) from the endomembrane system in the plane of cell division. PM: plasma membrane. In (ii), the vesicles have fused to form the cell plate which, as yet, has not become joined to the cell walls of the parent cell. The phragmoplast is beginning to lose definition. Finally, in (iii) the cell plate has joined onto the older cell wall of the parent cell and a primary wall is now evident. PL, plasmodesmata; ML, middle lamella. (b) TEM of a dividing cell from the base of the leaf sheath of *Echinochloa colonum* (L.), showing cell-plate formation within the phragmoplast. Abbreviations as for (a). Bar, 1 μm.

4.1.2 Primary wall formation

Once the cell plate has fused with the pre-existing walls of the parent cell, further wall material is laid down upon it to form the primary wall. The material derived from the cell plate becomes the middle lamella of the new wall, sandwiched between two layers of primary wall, one contributed by each daughter cell. When the primary wall has reached a certain thickness, generally $0.1–1.0 \, \mu m$, deposition of wall material continues at a rate sufficient to maintain approximately the same thickness as the cell grows. Once a particular part of the wall has ceased to increase in area, net primary wall synthesis generally stops, though some turnover of wall material may continue.

4.1.3 Secondary wall formation

Many cells contain only primary walls, and hence net wall synthesis ceases when the cell stops growing. However, some specialized cells (e.g. fibres, tracheids and vessels) continue to lay down further wall material over part or all of their surface, so that secondary wall is formed. Secondary wall is both thicker and more rigid than primary wall and is thought to be inextensible. Hence its deposition is probably sufficient to prevent further cell growth, except in the case of annular and spiral secondary walls (Chapter 5). Secondary wall formation is likely to be accompanied by changes in the composition of the middle lamella and the primary wall, such as lignification.

4.2 Biochemical pathways of wall polysaccharide formation

As far as we know, the substrates for the biosynthesis of all wall polysaccharides belong to the class of compounds known as **sugar nucleotides**, or nucleoside diphosphate sugars. These compounds contain a monosaccharide attached through its glycosidic hydroxyl to the β-phosphate of a ribonucleoside diphosphate (Figure 4.2 and Table 4.1). The base in the nucleoside is usually guanine or uracil. While these compounds are the only known substrates for wall polysaccharide synthesis, it is possible that other substrates might be found for synthetic systems that have yet to be elucidated, since some cytoplasmic polysaccharides are made from other types of substrate (e.g. fructan synthesis from sucrose).

4.2.1 Formation of sugar nucleotides

The sugar nucleotides are formed by the pathways shown in Figure 4.3.

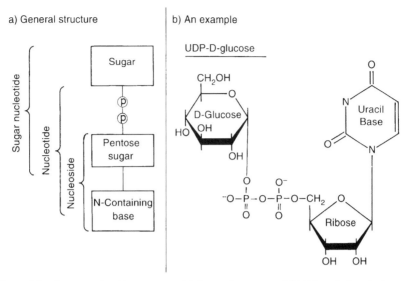

a) General structure

b) An example

Figure 4.2 Sugar nucleotide structure. (a) General structure. (b) Uridine diphosphate glucose (UDP–D–Glc).

Table 4.1 Sugar nucleotides found in plants

Name	Usual abbreviation
adenosine-diphosphate-L-arabinose	ADP-Ara
adenosine-diphosphate-D-fructose	ADP-Fru
adenosine-diphosphate-D-galactose	ADP-Gal
adenosine-diphosphate-D-glucose	ADP-Glc or ADPG
adenosine-diphosphate-D-mannose	ADP-Man
guanosine-diphosphate-D-fucose	GDP-Fuc
guanosine-diphosphate-D-galactose	GDP-Gal
guanosine-diphosphate-D-glucose	GDP-Glc or GDPG
guanosine-diphosphate-D-mannose	GDP-Man or GDPM
guanosine-diphosphate-L-rhamnose	GDP-Rha
thymidine-diphosphate-D-galactose	TDP-Gal
thymidine-diphosphate-D-galacturonic acid	TDP-GalA or TDP-GalU
thymidine-diphosphate-D-glucose	TDP-Glc or TDPG
thymidine-diphosphate-D-rhamnose	TDP-Rha
uridine-diphosphate-D-apiose	UDP-Api
uridine-diphosphate-L-arabinose	UDP-Ara
uridine-diphosphate-D-fructose	UDP-Fru
uridine-diphosphate-D-galactose	UDP-Gal
uridine-diphosphate-D-galacturonic acid	UDP-GalA or UDP-GalU
uridine-diphosphate-D-glucose	UDP-Glc or UDPG
uridine-diphosphate-D-glucuronic acid	UDP-GlcA or UDP-GlcU
uridine-diphosphate-N-acetyl-galactosamine	UDP-GalNAc
uridine-diphosphate-N-acetyl-glucosamine	UDP-GlcNAc
uridine-diphosphate-L-rhamnose	UDP-Rha
uridine-diphosphate-D-xylose	UDP-Xyl

The main flow of material is through UDP-Glc, which can be formed either from Glc-1-P or from sucrose. Other uridine-containing sugar nucleotides can be formed from UDP-Glc. There is also a separate pathway via myo-inositol, leading to UDP-GlcA. The guanine-containing sugar nucleotides are formed by a different set of reactions, in which GDP-Man is formed from Man-1-P and gives rise to other GDP-sugars. The enzymes involved in the early stages of these pathways are either free in the cytosol or loosely bound to the cytosol side of one of the cell's internal membranes. Some of the sugar nucleotide interconversion may occur within the lumen of the Golgi cisternae.

4.2.2 Polymerization: polysaccharide synthases

Formation of polysaccharide chains is carried out by membrane-bound enzymes which transfer monosaccharides from sugar nucleotides to growing polysaccharide chains, usually at the non-reducing end. The reactions are of the form:

$$\text{nucleoside diphosphate sugar} + \text{polysaccharide}_{n}$$
$$\downarrow$$
$$\text{nucleoside diphosphate} + \text{polysaccharide}_{(n+1)}$$

where n and $(n+1)$ indicate the number of sugar residues in the chain. The systematic names of these enzymes are of the form 'sugar-nucleotide-polysaccharide glycosyltransferase', but they are commonly called **polysaccharide synthases** or **synthetases**. The equilibrium of this type of reaction is strongly to the right, due to the breakage of a high-energy bond in the sugar nucleotide. Investigations into these enzymes have generally used impure membrane preparations, since the enzymes are difficult to solubilize and none have yet been purified. The experiments have involved the transfer of radioactive sugar residues from exogenously supplied radioactive sugar nucleotides to polysaccharide chains already present within the membrane preparation. The nature of these polysaccharides has seldom been investigated thoroughly, and hence there is often doubt as to the precise nature of the product formed. The enzymology of cell wall polysaccharide synthesis is therefore not yet fully explored. Table 4.2 gives an indication of the enzymes that have been found. The next two sections summarize our current knowledge of the roles these enzymes play in wall polysaccharide synthesis.

4.2.3 Polymerization: glucan synthesis

The major outstanding problem concerning the polysaccharide synthetases is that cellulose synthase has yet to be identified in vitro (Delmer et

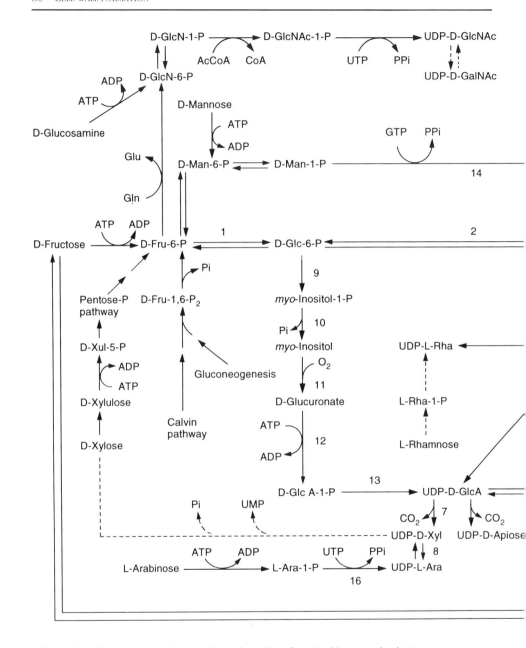

Figure 4.3   The known reactions leading to formation of nucleotide sugars in plants, as depicted by Feingold and Avigad (1980). Not all reactions have been demonstrated in the same tissue. Broken lines indicate reactions for which available evidence is fragmentary or preliminary. Arrows are drawn to show the direction in which the reaction is most likely to proceed under normal physiological conditions. Free monosaccharides are mostly the products of hydrolytic digestion of complex glycosides or sugar phosphates. (See abbreviation list and Table 4.1 for full identification.) Enzyme names: 1, hexose-phosphate

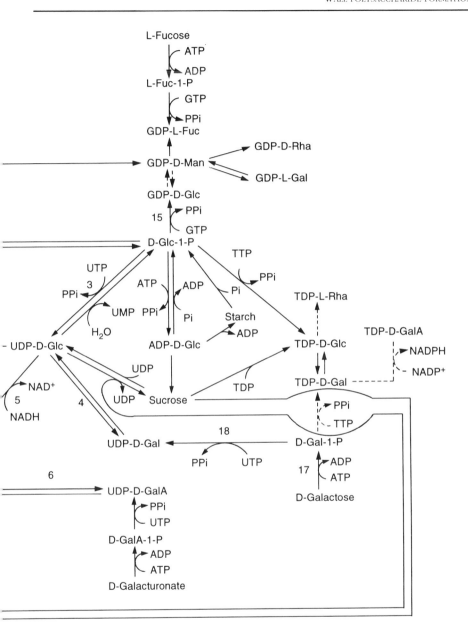

isomerase; 2, phosphoglucomutase; 3, UDP-D-glucose pyrophosphorylase; 4, UDP-D-glucose 4-epimerase; 5, UDP-D-glucose dehydrogenase; 6, UDP-D-glucuronate 4-epimerase; 7, UDP-D-glucuronate decarboxylase; 8, UDP-D-xylose 4-epimerase; 9, myo-inositol-1-phosphate synthase; 10, myo-inositol-1-phosphatase; 11, myo-inositol oxygenase; 12, D-glucuronokinase; 13, UDP-D-glucuronate pyrophosphorylase; 14, GDP-D-mannose pyrophosphorylase; 15, GDP-D-glucose pyrophosphorylase; 16, UDP-L-arabinose pyrophosphorylase; 17, D-galactokinase; 18, UDP-D-galactose pyrophosphorylase.

Table 4.2 Glycosyltransferases which form cell wall polysaccharides in higher plants

Sugar	Sugar-nucleotide substrate	Polysaccharide formed	Plant tissue	Remarks
glucose	UDP-glucose	β(1–4) glucan (possibly cellulose)	many, including cotton, mung bean, pea	may be difficult to distinguish from the enzymes which form xyloglucan and β(1–3), β(1–4) glucan
	UDP-glucose	β(1–3) glucan (callose)	many, including cotton, mung bean, pea	especially active at high (millimolar) UDP-Glc concentration
	UDP-glucose	β(1–3), β(1–4) glucan	cereals, e.g. oats, wheat	
	UDP-glucose	xyloglucan	pea seedlings, mung bean seedlings	may require UDP-Xyl to be present
	GDP-glucose	glucomannan (and possibly β(1–4) glucan)	mung bean seedlings, pea seedlings	sustained glucomannan formation requires GDP-mannose to be present
galactose	UDP-galactose	β(1–4) galactan	mung bean seedlings	
	UDP-galactose	galactomannan	developing fenugreek, guar and senna seeds	requires GDP-Man to be present
	UDP-galactose	xyloglucan	pea seedlings	
mannose	GDP-mannose	β(1–4) mannan and/or glucomannan	mung bean seedlings, pea seedlings	
	GDP-mannose	galactomannan	developing fenugreek, guar and senna seeds	
xylose	UDP-xylose	β(1–4) xylan	sycamore xylem, corn cobs, bean suspension culture	
	UDP-xylose	xyloglucan	pea seedlings, mung bean seedlings	requires UDP-Glc for sustained synthesis
arabinose	UDP-arabinose	arabinan and/or arabinogalactan	mung bean seedlings, bean suspension culture	
glucuronic acid	UDP-GlcA	glucuronoxylan	pea seedlings	requires UDP-Xyl for sustained synthesis
galacturonic acid	UDP-GalA	α(1–4) galacturonan	mung bean seedlings, pea seedlings, developing sycamore xylem	product may be homogalacturonan or part of RGI
fucose	GDP-Fuc	xyloglucan	pea seedlings	

al., 1993; Brown et al., 1994). There is strong evidence from in vivo pulse-chase studies in cotton fibres that the soluble precursor of cellulose is UDP-Glc, at least in that tissue. However, no glucosyltransferase has been obtained in vitro which forms $\beta(1-4)$glucan chains fast enough to account for the in vivo rate of cellulose synthesis. Furthermore, those enzymes which have been shown to give rise to $\beta(1-4)$-linked glucose in polysaccharide chains may not be involved in cellulose synthesis, since glucomannan and xyloglucan also contain this linkage, so it may be that cellulose synthase is inactivated when plant tissues are homogenized. Electron microscope studies suggest that cellulose microfibrils may be formed by a large, multienzyme complex which elongates the entire microfibril (section 4.5), and cytoskeletal elements may also be involved, so the system may be very fragile from the biochemist's point of view. It has been suggested that mild disruption of the cellulose synthase system may cause $\beta(1-3)$glucan formation (Chapter 7), and certainly it is very easy to demonstrate $\beta(1-3)$glucan synthesis in vitro.

Considerable research has been carried out on $\beta(1-3)$glucan synthesis in vitro, partly because of the importance of callose itself (Chapter 8) and partly in the hope that it will provide clues about cellulose synthesis. Solubilization of callose synthase in mild detergents indicates that a large enzyme complex is involved, with a mass of 450 kDa. This is thought to contain several different polypeptides. The use of 5-azido-UDP-Glc has permitted the identification of subunits which bind UDP-Glc. This has highlighted a 52–57 kDa polypeptide as a likely callose synthase component in several species. Callose synthase is stimulated by Ca^{2+}, polycations and β-glucosides, β-furfuryl-β-glucoside being a likely glucoside activator in vivo. Stimulation by Ca^{2+} may be due to the dissociation from the complex, when callose is present, of a 34 kDa polypeptide, which resembles animal annexins and which inhibits callose synthesis. A further group of polypeptides of 27–31 kDa may be important catalytic subunits, as judged by partial purification experiments.

A different $\beta(1-3)$glucan synthase is involved in callose formation in pollen tubes (Schupmann et al., 1994). Here, callose is a normal wall component (section 8.4). This activity is independent of calcium concentration, though it is activated by β-glucosides. A further, distinct glucan synthase is involved in $\beta(1-3),\beta(1-4)$glucan synthesis in monocots. This enzyme is also independent of calcium, but is stimulated by Mg^{2+} and Mn^{2+}.

Considerable knowledge has thus been obtained about the enzymes which form non-cellulosic β-glucans. So far this work has not led to any definite information about cellulose biosynthesis. An alternative approach involves seeking higher plant homologues for the bacterial cellulose synthases, which by contrast are highly active in vitro. Genes for these bacterial enzymes have been sequenced and possible homologous plant genes are being investigated in a number of laboratories.

4.2.4 Polymerization: heteropolysaccharide synthesis

In comparison with cellulose, the synthesis of the matrix polysaccharides is relatively easy to study. However, major areas remain to be explored. Most of these polysaccharides contain more than one type of monosaccharide, and some contain non-sugar components also (Chapter 2). The specificity of the biosynthetic enzymes is such that a different enzyme is needed to add each type of monosaccharide or other component to the growing polysaccharide chain. It is an open question at present as to whether these enzymes interact in a precisely controlled manner to produce a precise structure in the product, or whether they are only loosely controlled and hence introduce a degree of randomness into the polysaccharide (Figure 4.4). The former situation may apply in polysaccharides with a well-defined repeating unit, such as xyloglucan. The latter mechanism may apply in polysaccharides which have a less ordered arrangement of side-chains along the backbone.

The wall heteropolysaccharide whose synthesis is understood best is probably galactomannan (Edwards et al., 1992; section 2.6.3). This acts as a storage polymer in some legume seeds (section 8.5). The mannan backbone is elongated by a mannan synthase, whose rate of reaction is independent of the rate of addition of galactose side-chains. Thus a pure mannan is formed in vitro if the galactose donor, UDP-Gal, is absent. However, in the presence of UDP-Gal, galactose side-chains are added to the most recently attached mannose very shortly after the addition of the latter to the chain. Galactose cannot be added to a pre-formed mannan chain. This suggests that the two transferases form a complex which acts at the non-reducing end of the chain (Figure 4.4c). Not all mannose residues receive side-chain galactoses; the probability of galactose addition depends both on the concentration of UDP-Gal and on whether either or both of the nearest two mannoses in the chain have received side-chains.

Glucuronoarabinoxylan (GAX) synthesis is less well understood but shows similarities to galactomannan synthesis. The xylan backbone can be formed without addition of glucuronic acid side-chains. Xylosyltransferases of 38 and 40 kDa have been partially purified from French bean (Rodgers and Bolwell, 1992). In the presence of UDP-GlcA, glucuronic acid side-chains are added within a very short time of addition of xylose to the chain (Baydoun et al., 1989). When addition of arabinose and acetyl units occurs is not known. Methyl ether groups are added later, to the preformed polysaccharide, S-adenosyl-methionine being the donor.

Xyloglucan synthesis has received considerable attention but is only partly understood, in spite of its relatively regular structure (Maclachlan et al., 1992; White et al., 1993). Addition of xylose and glucose is proba-

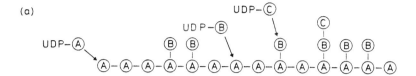

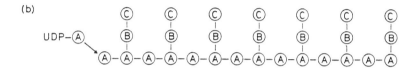

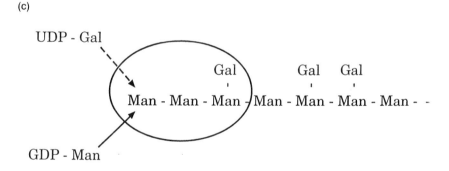

Figure 4.4 Heteropolysaccharides undergoing biosynthesis by (a) imprecise and (b) precise mechanisms. A, B and C represent different monosaccharides. Arrows represent glycosyl transfers in progress at one particular moment; UDP-sugars are depicted, since they are the most common sugar-nucleotide precursors, but this should not be taken to imply either that all the precursors are uridine derivatives or that no intermediates such as polyprenyl-phosphate sugars are involved in the glycosyl transfer. The overall direction of synthesis of the backbone is from right to left. (c) Schematic representation of galactomannan synthesis by a hypothetical synthase complex. The complex interacts with at least the three most recently added mannose residues and any galactose residue attached to them. The probability that a galactose is added to the most recently added mannose depends on the presence or absence of galactoses on the next two mannoses. Whether or not galactose is added, the chain is then elongated by the addition of another mannose.

bly tightly coupled, in a manner similar to galactomannan formation. Galactose and fucose appear to be added later, and can be added to the preformed polymer.

In glucomannan synthesis, the backbone is formed by addition of two types of monosaccharide. A single enzyme complex adds both mannose and glucose. It is possible that both sugars are added using only one binding site for the two very similar substrates, GDP-Glc and GDP-Man (Piro et al., 1993).

Very little is known about the enzymes which synthesize pectin. A galacturonyltransferase has been identified and is thought to form homogalacturonan. An arabinosyltransferase from French bean has been partially purified, and has a molecular weight of 70 kDa. Methyl ester groups are added from S-adenosyl-methionine.

4.2.5 Post-polymerization modifications

Very little is known about what, if any, modifications are made to the covalent structure of cell wall polysaccharides subsequent to their initial polymerization. Any covalent linkages that exist between different poly-saccharides may be formed after deposition of the polysaccharides in the wall, and links between polysaccharides and lignin are formed within the wall. Methyl groups are added to galacturonans and glu-curonoxylans after polymerization but while the polymers are still in the endomembrane system, prior to deposition in the wall. Addition of acetyl groups and ferulic acid is also thought to occur prior to deposi-tion; neither the substrates nor the enzymes have been identified, though acetyl-CoA and feruloyl-CoA have been suggested as substrates. There is evidence that large arabinogalactan side-chains become cova-lently linked to galacturonorhamnans within the wall as cells age. It is also known that some turnover of wall sugars occurs during cell growth (Chapter 5) and this may involve changes to the structure of some poly-saccharides. In the endosperms of developing seeds of *Senna occiden-talis*, the galactomannan which is deposited in the cell wall as a food reserve is modified after deposition by the removal of some of the galac-tose side-chains by the action of an α-galactosidase as the seeds approach maturity.

4.2.6 Possible involvement of glycolipid intermediates

In bacteria, the extracellular polysaccharides and glycoproteins are often formed via glycolipid intermediates. Sugars are passed from sugar nucleotides either directly onto the phosphate group of a polyprenol phosphate, or onto a short oligosaccharide attached to such a lipid. The mono- and oligosaccharides are then transferred onto the growing poly-saccharide or glycoprotein.

Considerable effort has been devoted to finding out whether a similar situation exists in higher plants. However, no good evidence has been obtained for this as far as cell wall polysaccharides are concerned. Polyprenol phosphate sugars are found in plants but their role appears to be in the biosynthesis of glycoproteins rather than polysaccharides.

4.2.7 Priming systems for polysaccharide biosynthesis

The polysaccharide synthases which add sugar residues to nascent matrix polysaccharides require a pre-existing polysaccharide on to which to initiate the synthesis of the polysaccharide. These enzymes might simply transfer one sugar residue from a sugar nucleotide onto

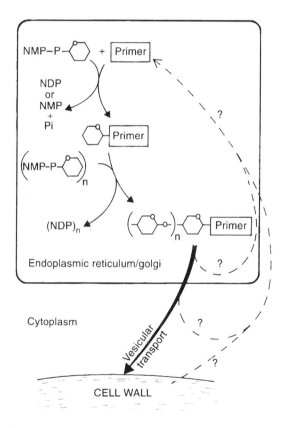

Figure 4.5   Possible mechanism of primer involvement in the synthesis of non-cellulosic polysaccharides. In this scenario, the primer may be detached from the polysaccharide during or after synthesis, possibly after insertion of the polysaccharide into the cell wall, or it may be retained on the polysaccharide and function in the structure of the wall. NMP, nucleoside monophosphate; NDP, nucleoside diphosphate.

another, free, sugar residue to make the first glycosidic bond of the polysaccharide. However, comparison with the biosynthesis of other polysaccharides suggests that the first (reducing end) sugar is likely to be attached initially to an aglycone primer. The glycosylated primer will then act as an acceptor for the transfer of further sugar residues as the polysaccharide begins to elongate (Figure 4.5). In a number of polysaccharides, such as starch and glycogen, the aglycone is a protein, and there is evidence for the participation of protein primers in the biosynthesis of glucuronoxylan in peas (Crosthwaite et al., 1994) and of xyloglucan in suspension-cultured bean cells. There is also evidence for the involvement of inositol as a primer in callose (β(1–3)glucan) biosynthesis. It is not known whether the primers are cleaved from the polysaccharide before its incorporation into the cell wall, or whether the primers are themselves incorporated into the structure of the cell wall.

4.3 Biochemical pathways of wall protein formation

Cell wall proteins are produced by the endomembrane system and exported via the Golgi apparatus to the cell wall. The polypeptides are formed by the ribosomes of the rough endoplasmic reticulum. In the case of hydroxyproline-rich glycoproteins (HRGPs), and other glycoproteins which contain lower amounts of hydroxyproline, some or all of the proline residues are hydroxylated, probably after completion of the polypeptide chain. The hydroxylation is carried out by the enzyme, prolyl hydroxylase. The enzyme uses molecular oxygen as the oxidizing agent, and requires α-ketoglutarate as a co-substrate and ascorbate and ferrous ions as co-factors (Figure 4.6). The enzyme shows highest activity when acting on a series of four or more adjacent proline residues in a polypeptide chain. This is due to its recognition of the polyproline II helical structure taken up by such contiguous proline residues in HRGP molecules (section 2.7).

Addition of arabinose to hydroxyproline occurs by transfer from UDP-Ara. At least two different enzymes are thought to be involved, given the presence of both α- and β-linked arabinose residues in the oligosaccharide side-chains. Nothing is known about the mechanism of galactose addition to serine, though analogy with other systems would suggest that the donor is UDP-Gal.

Not all prolines within HRGPs are hydroxylated, and not all hydroxyprolines are glycosylated. The amino acid sequences which control these events are being explored (Kieliszewski and Lamport, 1994). In particular, contiguous hydroxyprolines are much more likely to be arabinosylated than single ones.

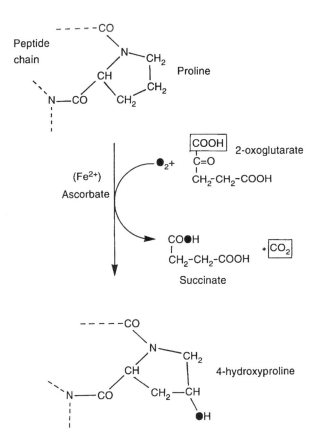

Figure 4.6   The hydroxylation of poly-L-proline by prolyl hydroxylase: a mixed function oxygenase (prolyl peptide, 2-oxoglutarate: oxygen oxidoreductase, EC 1.14.11.2)

4.4 Biochemical pathways of lignin formation

Lignin is formed from three aromatic alcohols: coumaryl, coniferyl and sinapyl alcohols. The alcohols are formed by pathways leading from the aromatic amino acids, phenylalanine and tyrosine. The major pathway leads from phenylalanine through the enzyme, phenylalanine-ammonia lyase (PAL) (Figure 4.7). The alcohols are oxidized within the wall by the action of peroxidase, and perhaps also by polyphenol oxidases such as laccase, giving rise to mesomeric phenoxy radicals (Figure 4.8). These

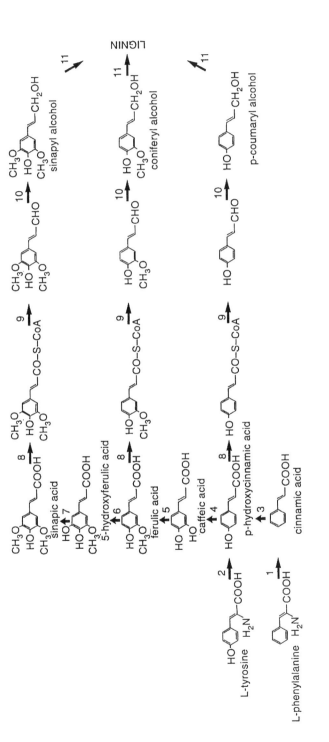

Figure 4.7 Pathway of reactions involved in lignin biosynthesis. Enzyme names: 1, phenylalanine-ammonia lyase (PAL); 2, tyrosine-ammonia lyase (TAL); 3, cinnamic acid 4-hydroxylase (CA4H); 4, coumarate 3-hydroxylase; 5, SAM:caffeate 3-O-methyl transferase (COMT); 6, ferulate 5-hydroxylase; 7, 5-hydroxyferulate O-methyl transferase; 8, 4-coumarate:CoA ligase (4CL); 9, cinnamoyl-CoA:NADPH oxidoreductase; 10, cinnamyl alcohol: NADP oxidoreductase (CAD); 11, peroxidase and laccase.

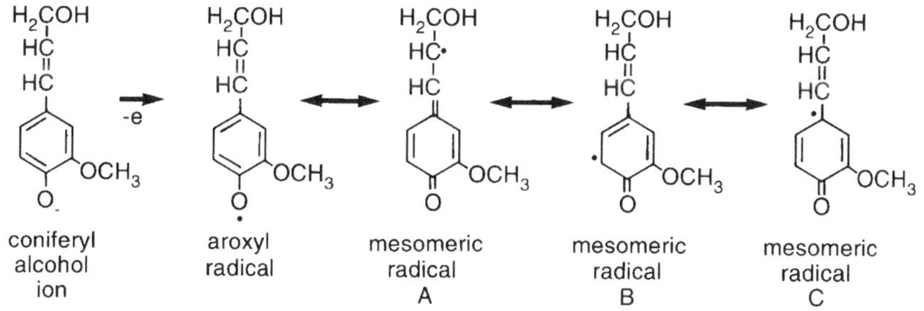

Figure 4.8 Examples of mesomeric phenoxy radicals produced from coniferyl alcohol, formed as a result of peroxidase activity.

radicals react together spontaneously to form lignin (Figure 4.14). They probably also react spontaneously with some polysaccharides to give linkages between lignin and wall polysaccharides.

4.5 Sites of formation of cell wall polymers

4.5.1 Polysaccharides

The sugar nucleotides which are substrates for polysaccharide biosynthesis are formed by enzymes which are located in the cytosol. However, the glycosyltransferases which carry out the polymerization reactions are, as far as is known, all membrane-bound. Considerable work has been carried out on the identification of the membranous organelles in which the polysaccharide synthases are located.

Early work involved autoradiography of plant tissues which had been supplied with a probe of radioactive sucrose or glucose, often followed by a chase of non-radioactive precursors. Parallel biochemical studies were carried out to identify the polysaccharides that had been formed. Using a variety of different tissues and experimental conditions, these studies indicated that newly synthesized pectin and hemicellulose could be found located in the endoplasmic reticulum, Golgi apparatus and Golgi vesicles, as well as in the cell wall, while newly synthesized glucans were only found at the plasma membrane and within the cell wall. The conclusions were strengthened by investigations using cell fractionation techniques. In these studies, plant tissues (generally root-tips) were fed with radioactive precursors (glucose or sucrose) and the tissue was then homogenized in the presence of a fixative, glutaraldehyde, which helped to preserve organelle structure. The homogenate was layered on

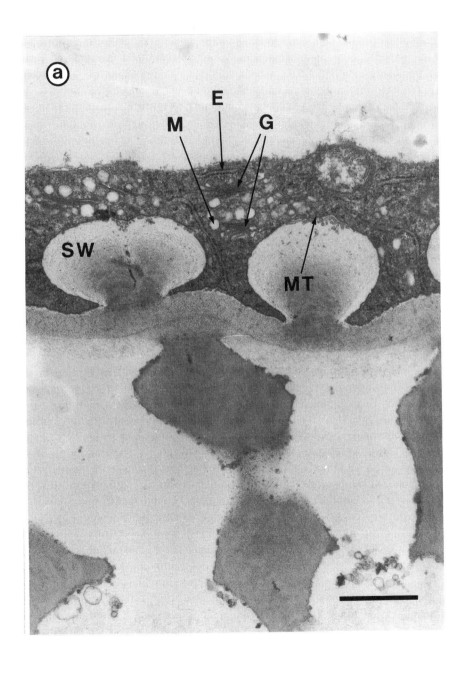

(b)

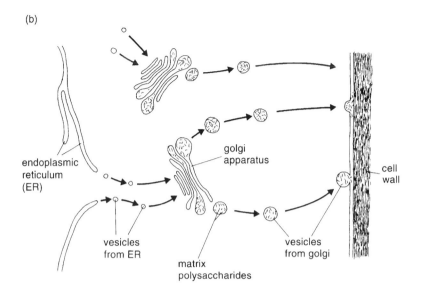

(c)

Figure 4.9 (a) TEM of the endomembrane system in young xylem (LS) in the papery pods of *Sutherlandia frutescens* (Balloon pea). G, Golgi apparatus; M, matrix polysaccharides; E, endoplasmic reticulum; SW, secondary wall; MT, microtubules. Bar, 1 μm. (b) Line drawing showing the movement of cell wall matrix polysaccharide from the endomembrane system to the cell wall via Golgi-derived vesicles. As discussed in the text, the majority of matrix polysaccharide biosynthesis occurs in the Golgi apparatus. Nevertheless, there is evidence that some may take place in the endoplasmic reticulum. (c) Line drawing of the subcompartments of the Golgi apparatus. The number of cisternae in the cis, medial and trans subcompartments is variable; here three are shown in each of these subcompartments, with one in the trans-Golgi network (TGN).

top of a discontinuous sucrose-density gradient, and then centrifuged in order to obtain fractions enriched in cell walls, Golgi apparatus and endoplasmic reticulum. The cell-wall-rich fractions (which probably also contained substantial amounts of plasma membrane) contained radioactive sugars characteristic of all the wall polysaccharides, while the Golgi- and endoplasmic-reticulum-rich fraction contained a relatively low proportion of glucose.

In more recent work, the subcellular distribution of polysaccharide synthases has been investigated by sucrose density-gradient fractionation of unfixed homogenates. Enzyme markers have been used to identify the various organelles. With one exception, the glycosyltransferases examined have been found in the Golgi apparatus (Figure 4.9). These include an arabinosyltransferase, a xylosyltransferase involved in $\beta(1–4)$xylan synthesis, a xylosyltransferase and a glucosyltransferase which synthesize xyloglucan, a glucuronyltransferase involved in glucuronoxylan synthesis, and one or more glucosyltransferases which operate at low (micromolar) concentrations of UDP-Glc to produce both $\beta(1–3)$- and $\beta(1–4)$-linked glucans (glucan synthase I). Of these, the arabinosyltransferase is also thought to be located partially in the endoplasmic reticulum. The one polysaccharide synthetase that has been found to be absent from the Golgi apparatus but present in another organelle is a glucan synthase activity, glucan synthase II, which is activated by its substrate (UDP-Glc) and is thus detected best at high (millimolar) substrate concentrations. This enzyme activity, which forms predominantly $\alpha(1–3)$glucan, is associated with the plasma membrane. It should be noted that the terms glucan synthase I and II, often abbreviated to GSI and GSII, do not represent well-characterized, individual enzymes. They are used to denote two sets of assay conditions, and the activity seen in membrane fractions is likely to involve several different enzymes, especially in the case of GSI.

The conclusion from these studies is that the matrix polysaccharides are formed in the endomembrane system of the cell. The major part of the activity occurs in the Golgi apparatus, but early stages, perhaps including any priming reactions, may occur in the endoplasmic reticulum. Those tissues which show the greatest amount of cell-wall-like polysaccharide accumulating in the endoplasmic reticulum are those, such as maize root tips, which secrete large amounts of mucilage. This mucilage contains glycoproteins, of which the polysaccharide part closely resembles pectin, and hence it may be that the synthesis of such glycoproteins occurs to a large degree in the endoplasmic reticulum. Even in this case, however, it is expected that the material will be secreted via the Golgi apparatus.

In higher plants, there is no evidence for the synthesis of cellulose, or of any polymeric precursors of it, in the endomembrane system. No microfibrils are seen inside the plasma membrane in the electron microscope, and such $\beta(1-4)$glucan synthase activity as is found in the Golgi apparatus may be attributed to the formation of the glucan backbone of xyloglucan, and perhaps also to glucomannan synthesis. The mucilage of maize root tips also contains some $\beta(1-4)$ glucan, forming part of the complex glycoprotein. In certain chrysophycean algae, the Golgi apparatus synthesizes scales which contain substantial amounts of $\beta(1-4)$ glucan, but here again the glucan may be thought of as part of a larger complex, rather than as true cellulose. The available evidence suggests that cellulose is formed at or outside the plasma membrane. Groups, or 'rosettes', of particles are seen in the plasma membrane when viewed by freeze–fracture techniques (Figure 4.10) and these rosettes are sometimes associated with the ends of microfibrils (Figure 4.11). They are thought to be 'cellulose synthase' complexes, involved in the elongation of whole microfibrils (Figure 4.12). This means that all the chains in one microfibril would need to be elongated at the same rate by the complex, a requirement that means that the complex would need to be formed of many subunits, each elongating one chain at a time. Since there are between 30 and 100 chains in the cross-section of one microfibril, this process would be one of remarkable complexity, but nevertheless the evidence favours it.

One important piece of evidence is that the crystalline structure of cellulose, the cellulose I structure, is not the lowest energy state available (section 2.2). It is unlikely, therefore, to be formed by crystallization of preformed cellulose chains, since such a process would give rise to the cellulose II structure. Hence a process in which crystallization accompanies, or follows very closely upon, addition of glucose residues to the chains seems probable; such a process would permit the geometry of the enzyme complex to govern the crystalline form of the microfibril. The size of the putative cellulose synthase complex is so great that it would be expected to be easily disrupted by tissue homogenization. This could account for the failure to observe cellulose synthesis in vitro.

It has been suggested that the disrupted complex might synthesize callose, rather than cellulose. This is an attractive suggestion, since it would account for the considerable variations seen in the proportion of 1,4 links to 1,3 links formed by glucan synthase I under different conditions (and even when the conditions are kept as constant as possible). It would also provide a neat explanation for the extraordinarily high activity of $\beta(1-3)$glucan synthase which appears almost instantly when plant tissues are wounded. However, there is no direct evidence for this proposal.

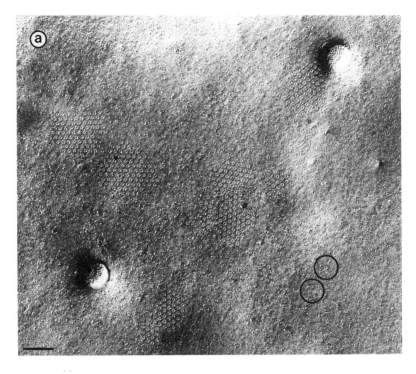

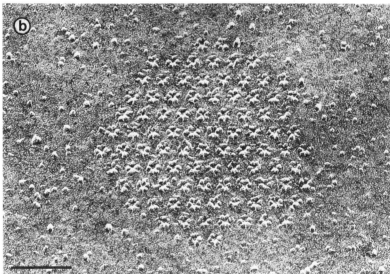

Figure 4.10 (a) The P-face of the plasma membrane of *Micrasterias denticulata* engaged in the synthesis of the secondary wall. Hexagonal arrays of rosettes of varying size and shape, as well as a few isolated rosettes (upper circle) can be observed. The lower circle indicates an array of three rosettes. The large circular indentations may correspond to forming slime secretion pore complexes. Bar, 0.2 μm. (b) The P-face of a complex consisting of a hexagonal array of rosettes at high magnification. The best-preserved rosettes are composed of six particles. Bar, 0.1 μm.

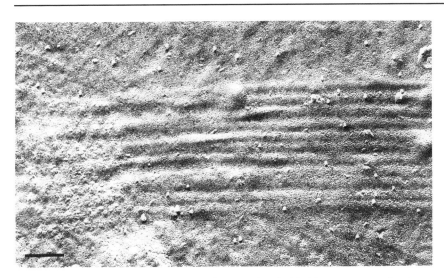

Figure 4.11 The E-face of a rosette complex associated with ridges corresponding to newly synthesized fibrils. Bar, 0.1 μm. Tissue as in Figure 4.10.

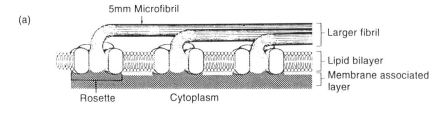

(a)

5mm Microfibril

Larger fibril

Lipid bilayer

Membrane associated layer

Rosette   Cytoplasm

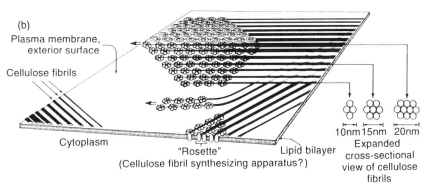

(b)

Plasma membrane, exterior surface

Cellulose fibrils

Cytoplasm

"Rosette" (Cellulose fibril synthesizing apparatus?)   Lipid bilayer

10nm 15nm 20nm
Expanded cross-sectional view of cellulose fibrils

Figure 4.12 Model of cellulose fibril deposition during secondary wall formation in *Micrasterias*. Each rosette is believed to form one 5-nm microfibril. A row of rosettes forms a set of 5-nm microfibrils, which aggregate laterally to form the larger fibrils of the secondary wall. (a) Side view. The stippled area in the centre of a rosette represents the presumptive site of microfibril formation, although details of its structure, composition and enzymic activity remain unclear. The 'membrane-associated layer' may serve to hold the rosettes together in the hexagonal array. (b) Surface view with expanded cross-sectional view of cellulose fibrils.

In most cells, cellulose appears to be laid down at the outer surface of the plasma membrane, and hence its synthesis could be accounted for by an enzyme associated with the plasma membrane. However, there are a few cases such as the parenchyma of oat coleoptiles, in which some new wall material appears to be laid down at the outer edge of the cell wall. At present, there are no indications as to how the cellulosic portion of this material is synthesized.

The polysaccharide synthases involved in cell wall biosynthesis all appear to utilize sugar-nucleotide substrates, which are formed mainly in the cytosol, but the products of the enzymes are separated from the cytosol by a membrane – either the endoplasmic reticulum, the Golgi apparatus membrane or the plasma membrane. In the case of the polysaccharide synthases of the endomembrane system, the enzymes are thought to be located on the luminal side of the membrane; the sugar-nucleotides are thought to be transported from the cytosol into the lumen by permeases in the membrane (Figure 4.13a). However, direct donation of sugars from the cytosol to the cytosolic face of transmembrane enzyme complexes is also a possibility (Figure 4.13b). The complexes would then need to transfer the sugar residues to the luminal face before or during the addition to the nascent polymer. In the case of the putative cellulose synthase complex, direct donation of the sugars to the complex at the cytosol side of the plasma membrane seems most likely, since transfer of the intact substrates to the outside of the plasma membrane would be expected to lead to considerable diffusion of the substance away from the complex, into the wall and extracellular space.

While most of the pathways for sugar-nucleotide formation occur in the cytosol, some may be formed within the lumen of the Golgi cisternae. Both UDP-GlcA dehydrogenase and UDP-Xyl 4-epimerase have been found there. This may be important for maintaining the availability of all three sugars needed for GAX synthesis.

4.5.2 Protein

The cell wall proteins, like other proteins destined for export beyond the plasma membrane, are synthesized by the ribosomes of the rough endoplasmic reticulum. The nascent polypeptide crosses the membrane and either remains attached to the luminal side of the membrane or is released into the lumen. Hydroxylation of proline residues (Figure 4.6) occurs both in the endoplasmic reticulum and in the Golgi apparatus. The arabinosyltransferases involved in the glycosylation of hydroxyproline residues also appear to be located in both these organelles, with the first arabinosylation perhaps occurring in the endoplasmic reticulum and subsequent arabinose residues being added in the Golgi apparatus.

The cell wall proteins are exported to the cell wall via Golgi vesicles, as in the case of the matrix polysaccharides. It is thought that cross-links between cell wall proteins are probably formed in the cell wall.

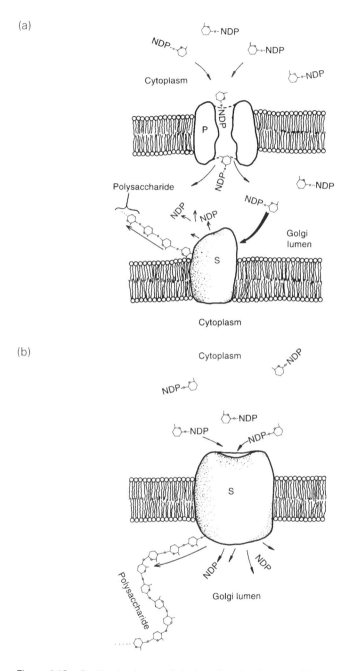

Figure 4.13 Synthesis of non-cellulosic cell-wall polysaccharides by the endomembrane system. Because the polysaccharides are synthesized in the lumen of the Golgi and endoplasmic reticulum, the sugar-nucleotide precursors must, at some point, cross the membrane boundary. This might occur via a translocating enzyme that is separate from the glycan synthase as shown in (a). Alternatively, the synthase may itself possess a translocating function which could deliver a sugar nucleotide from the cytoplasm directly onto the active site of the enzyme (b). S, synthase; P, permease.

4.5.3 Lignin

The polymerization of lignin occurs in the cell wall (Figure 4.14). It usually begins in the middle lamella and then spreads first to the primary wall and then to the secondary wall.

Lignin precursors are formed at or near the endoplasmic reticulum. Phenylalanine-ammonia lyase is found either free in the cytoplasm or bound to the cytoplasmic face of the ER. The enzymes which carry out the subsequent hydroxylation reactions are present in the ER, and the intermediates in the pathway, being quite lipophilic, may remain dissolved in the lipid of the membrane during the reactions. It is possible that the lignin precursors are then transported to the cell wall via the Golgi apparatus and Golgi vesicles, along with the matrix polysaccharides and proteins.

4.5.4 Organization of biosynthesis and transport within the Golgi apparatus

As noted in the previous sections, most of the biosynthetic events involved in matrix polysaccharide formation occur in the Golgi apparatus, and this organelle is also involved in the transport and modification of wall proteins and glycoproteins. The detailed organization of the Golgi is now being studied. The isolated organelle can be subfractionated into cis-, medial- and trans-Golgi subfractions by ultracentrifugation on shallow density gradients. Golgi vesicles can also be obtained in this way. In a complementary approach, immunocytochemistry can reveal the content of each subfraction, and a '**trans-Golgi network**' (TGN) can also be identified. These studies are providing information on the roles of each part of the Golgi apparatus.

Subfractionation suggests that the glucuronyl- and xylosyltransferases involved in GAX formation occur mainly in the cis-Golgi (Hobbs et al., 1991). Immunocytochemistry indicates that the backbone of RG I is formed in the cis- and medial-Golgi, with methylation occurring in the medial cisternae and addition of arabinose in the trans-cisternae (Lynch and Staehelin, 1992; Zhang and Staehelin, 1992). Both approaches suggest that the glucan chain of xyloglucan is formed in the trans-cisternae, with fucosylation occurring in the trans-cisternae and TGN. Immunocytochemical evidence indicates that the addition of arabinose to extensin occurs in the cis-Golgi, as does the trimming of the core oligosaccharide of N-linked glycoproteins. Subsequent further glycosylation of these oligosaccharides takes place in the medial and trans-cisternae and the TGN.

Transport of nascent wall material from the endoplasmic reticulum to the cis-Golgi, from one Golgi compartment to another and from the TGN to the plasma membrane all occur via membrane-bound vesicles. A

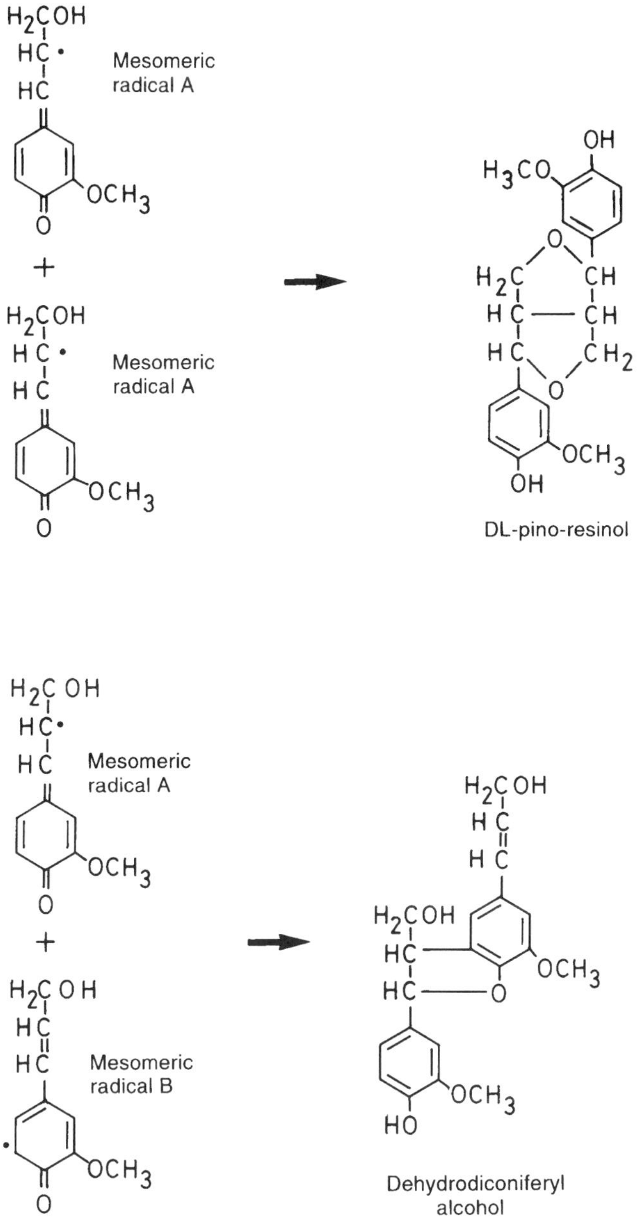

Figure 4.14   Typical events that occur during the polymerization of lignin in the cell wall from some of the radicals described in Figure 4.8.

vesicle buds from the donor compartment and then fuses with the membrane of the acceptor compartment, transferring both membrane and vesicle contents to the latter. It is probable that specific amino acid sequences ('**signal sequences**') within the primary structure control the location of proteins which are localized specifically in one compartment. The 'default' pathway is that which leads right through to the plasma membrane; thus, proteins move from the endoplasmic reticulum, through the Golgi subcompartments to the plasma membrane, unless they contain a signal sequence which causes them to be retained in one compartment along the way (or to be directed from the TGN to the tonoplast). The mechanisms by which vesicles are directed specifically to the correct target membrane are beginning to be worked out in yeast; in higher plants the mechanism is likely to be very similar, since it seems to be highly conserved within eucaryotic cells in general (Ferro-Novick and Jahn, 1994).

4.6 Control of wall formation

4.6.1 Control of microfibril orientation

There is now considerable evidence that the initial orientation of newly synthesized microfibrils is governed by the orientation of microtubules

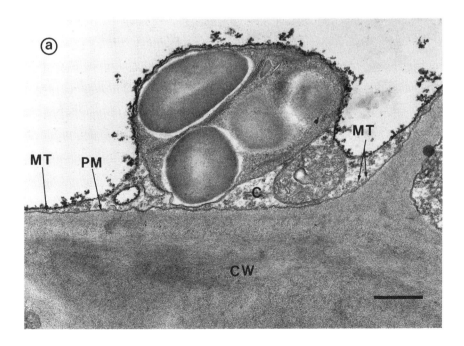

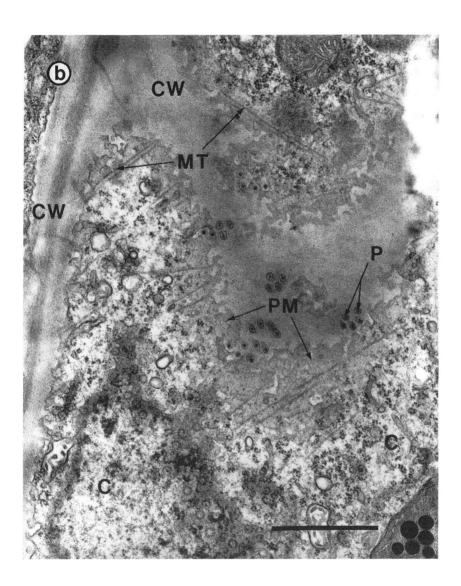

Figure 4.15   Microtubules adjacent to the plasma membrane in cells from the base of the leaf sheath in *Echinochloa colonum* (L.). (a) TEM showing section through both plasma membrane and associated microtubules. Bar, 1μm. (b) TEM of a section made in the plane of the plasma membrane. Bar, 500 nm. CW, cell wall; PM, plasma membrane; MT, microtubules; C, cytoplasm; P, plasmodesmata.

Figure 4.16   The relationship between
microtubules adjacent to the plasma
membrane, and microfibrils of the cell
wall. (a) Immunofluorescence light-
micrograph of onion root hair
microtubules which can be seen to lie at
45° to the long axis of the root hair. Note
that the root hair has been compressed so
that fluorescence from labelled
microtubules adjacent to both near and far
side of the root hair is in focus. In other
words, the microtubules follow a helical
path along the root hair. (b) TEM of freeze-
fractured replica of onion root-hair cell
wall. The microfibrils of the inner part of
the wall (IW) are at 45° to the long axis
(arrow) as are the microtubules adjacent
to the plasma membrane in (a), and
successive layers are at right-angles to
each other. The microfibrils of the outer
wall (OW), appear to have a random
texture.

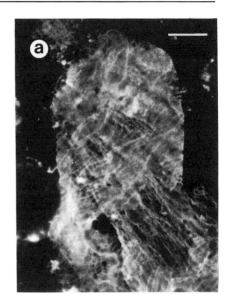

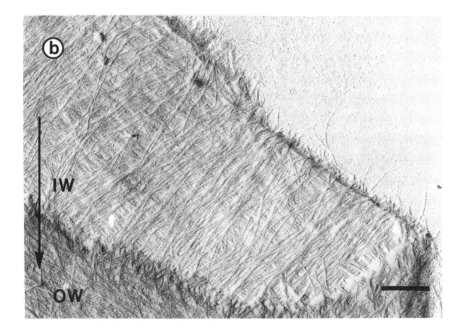

in the cytoplasm immediately beneath the plasma membrane (Figures 4.15 and 4.16; Giddings and Staehelin, 1991). The two systems usually lie parallel, and a causal relationship is indicated by the observation that agents, such as colchicine, which disrupt cytoplasmic microtubules also cause disorganization of the newly synthesized cell wall microfibrils (though cell wall synthesis continues, in a disorganized manner). The mechanism by which microtubules might govern microfibril orientation is uncertain. Since the microfibrils are thought to be synthesized by enzyme complexes in the plasma membrane, and the microtubules lie very close to the plasma membrane, a direct physical influence is possible. This influence would probably relate to the control of the direction of movement of the synthase complexes within the plane of the membrane. It is thought that since the cellulose microfibrils are very much larger than the postulated synthase complexes, each complex adds material to the end of a microfibril by moving in the plane of the membrane, spinning out the microfibril behind it, while the microfibril remains stationary. There is evidence from electron micrographs that the complexes may sometimes be aligned in rows parallel to the microtubules, and this may relate to the control of microfibril orientation. The orientation of microtubules may in turn be governed by the orientation of actin microfilaments, since microfilaments have a similar orientation to microtubules and their disruption by cytochalasin B leads to microtubule disorientation.

Well established though the theory of microtubular control of microfibril orientation is, it is not the whole story. As mentioned in section 3.5, some cell walls show a helicoidal pattern of orientation, which suggests a process of self-assembly, analogous to that found for helicoidal structures in animals. The control of such a process would probably reside in the matrix polymers, rather than the microfibrils, since microfibrils do not have sufficient potential for structural variation to account for the varying degrees of helicoidal twist found in different walls. Hemicelluloses have been found to form helicoidal patterns spontaneously in vitro, and hence are likely to be involved in the control of microfibril orientation in helicoidal walls (Reis et al., 1991). The self-assembly properties of cell walls are probably less important in the initial orientation of microfibrils than the effects of cortical microtubules. However, the two may reinforce each other and in those situations where microfibrils are absent, either naturally or due to experimental disaggregation, the self-assembly properties may take control.

Whatever the manner in which the initial microfibril orientation is determined, the microfibrils can subsequently be reorientated passively by stretching of the wall during cell enlargement. Since older parts of the wall have usually been subjected to more stretching than younger parts during cell extension, the older parts of the wall of a cell that is extending are usually more longitudinally orientated than younger parts. This passive reorientation of microfibrils was an important part of

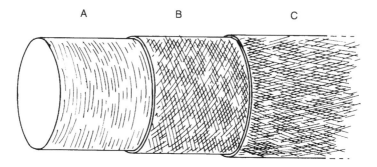

Figure 4.17 Multinet growth hypothesis of Roelofsen and Houwink (1953). Microfibrils, the directions of which are represented by shading, are laid down at the plasma membrane perpendicular to the direction of cell elongation (A). As the cell ages and elongates, the microfibrils become realigned in the direction of elongation (B and C).

the 'multinet growth' hypothesis of Roelofson (Figure 4.17), and appears to be a general property of growing plant cells.

4.6.2 Control of matrix deposition

The composition of the cell wall matrix varies considerably in different cell types and at different stages of wall deposition. In particular, there is usually a considerable difference between primary and secondary cell walls, and much work has been devoted to studying the control mechanisms which operate to bring about these different matrix compositions.

As far as the polysaccharides of the matrix are concerned, the main control appears to be at the level of the polysaccharide synthases. At the onset of secondary wall deposition in differentiating xylem tracheids in dicots, the activity of enzymes involved in pectin biosynthesis (galacturonyl- and arabinosyltransferases) falls, while there is a rise in the activity of the xylosyltransferase which forms xylan. In contrast, the activity of the arabinosyltransferase rises during cell division in non-differentiating cells. In the cases of both arabinosyl- and xylosyltransferases, increases in activity are dependent on transcription and translation, indicating that new mRNA and enzyme molecules must be synthesized. In *Pinus sylvestris*, a gymnosperm, xylem differentiation is accompanied by a large increase in the activity of the glucomannan synthase system.

A subsidiary level of control may operate at the level of the enzymes which form sugar-nucleotides. The activities of the two enzymes which together form UDP-xylose from UDP-glucose increase during secondary wall formation, though the epimerases which form the precursors of pectin continue to remain active. Several of the enzymes which form sugar-nucleotides vary in activity during the cell cycle in *Catharanthus roseus* cultures. Other possible control points include the transport of sugar-nucleotides from the cytosol to the lumen of the endoplasmic reticulum and the Golgi apparatus, and the fusion of the Golgi vesicles with the plasma membrane, which is a Ca^{2+}-dependent process.

There may be a degree of feedback control of flux through the polysaccharide biosynthesis pathway. For instance, xylan synthase in sycamore is inhibited by nucleoside mono- and diphosphates, while UDP-glucose dehydrogenase is inhibited by UDP-xylose.

The initiation of lignin formation in differentiating xylem involves increases in the activity of PAL and of other enzymes involved in the synthesis of the lignin precursors. For the initiation of PAL activity, both transcription and translation are necessary, so de novo synthesis of the enzyme is likely to occur. There is also some evidence for a decrease in the rate of PAL breakdown. In some cases, the activity of peroxidase in the cell wall has also been found to increase during the onset of lignification. Immunocytochemical studies of developing xylem have shown that peroxidase is expressed in the secondary thickenings of xylem tracheids where lignin is being deposited. PAL and coumarate-4-hydroxylase, the first two enzymes of the pathway supplying lignin precursors, are expressed in the xylem parenchyma, indicating a division of labour in the process of lignin formation between different cell types in the xylem.

Matrix deposition must also be controlled with respect to the sites in the cell surface at which new material is deposited. This is probably achieved by controlling the flow of vesicles from the trans-Golgi and TGN to the plasma membrane. The vesicles contain matrix polysaccharides and proteins, plus perhaps lignin precursors and peroxidase in lignifying cells, and also possibly the putative cellulase synthase complexes. The site of fusion with the plasma membrane varies greatly: during cytokinesis, it is the cell plate (Figure 4.1); during elongation growth it is the lateral walls; during secondary wall thickening it is beneath the thickening regions of the wall (Figure 4.9); in pollen tubes it is at the tube tips (Figure 8.5). How vesicle movement is controlled is not known; ATP-dependent movement along actin microfilaments or microtubules may be involved.

Once vesicles from the Golgi apparatus have fused with the plasma membrane and released their contents into the cell wall, the matrix polymers must form non-covalent or covalent bonds with the existing wall material in order to be integrated into the wall. Little is known about this insertion process. For xyloglucan, the backbone of newly synthesized molecules may become covalently bound to the backbone of pre-existing xyloglucan by the action of the enzyme, xyloglucan endo-transglycosylase (XET) (section 5.5.2.). Xyloglucan, xylans and gluco-mannans are thought to form hydrogen bonds with nascent cellulose, a process which may contribute to the control of microfibril orientation (section 4.6.1.; Reis et al., 1992). Extensin rapidly becomes insolubilized after secretion into the wall, perhaps due to the formation of isodityro-sine cross-links. In general, the process of wall assembly is an important area for research in the immediate future (Jarvis, 1992).

4.6.3. The influence of growth substances on cell wall deposition

Both cell enlargement and cell differentiation are influenced by growth substances, and hence both the rate of cell wall synthesis and the type of material incorporated are influenced by growth substances. Auxin, gibberellin and ethylene all influence cell growth and this is reflected in changes in the levels of activity of all the glysosyltransferases which synthesize wall polysaccharides. Changes in the relative amounts of growth substances can induce differentiation, for instance in the induction of xylem differentiation in bean tissue cultures by high NAA : kinetin ratios. This also brings about major changes in the activity of glycosyltrans-ferases and the enzymes of lignin biosynthesis. The pattern of wall deposition alters dramatically at the onset of xylogenesis, and this is probably mediated by rearrangements of the cytoskeleton (Shibaoka, 1994). Single-celled cultures of *Zinnia elegans* offer a particularly clear example of this and promise to be a powerful tool for elucidating the mechanisms by which growth substances bring about differentiation (Seagull and Falconer, 1991). However, neither for cell enlargement nor for cell differentiation are the mechanisms by which growth substances exert their effects yet known. The involvement of transcription and translation in the changes that occur during differentiation suggests that the control mechanisms include effects at the genome level, but how these are exerted is not understood.

4.6.4 Influence of external stimuli on wall formation

The deposition of cell wall material is sensitive to external stimuli as well as to internal signals. Many such stimuli present stressful challenges to the plant cell. For instance, wounding of plant cells brings about the rapid deposition of callose at the wound site. This may be by perturba-tion of the cellulose synthesis system (section 4.2). Callose formation may be triggered by an influx of calcium across the plasma membrane. The same ion flux may trigger action potentials in the plasma mem-brane, which may cause other wound responses. Such additional wound responses often include the deposition of phenolic and lipophilic wall material and HRGPs at the wound site, together with the production of ethylene and phytoalexins. Other stresses, such as osmotic and drought stresses, have similar effects; where water stress occurs, abscisic acid may play a part in the signal transduction system. High gravitational fields cause a general reduction in growth, combined with an alteration in wall phenolic composition (Waldron and Brett, 1990). Invasion by microorganisms triggers responses similar to those of wounding (Chapter 7), and light/dark transitions affect growth rates and hence rates of wall synthesis (Chapter 5).

Summary

The cell plate gives rise to the middle lamella, upon which the primary wall is deposited during the phase of cell growth. Secondary wall is subsequently laid down in some cells, once wall extension has ceased. The polysaccharides of the wall are synthesized from sugar nucleotides by polysaccharide synthases (glycosyltransferases), while lignin is formed from aromatic alcohols. Cellulose is formed by enzymes in the plasma membrane, in contrast to the matrix polysaccharides, proteins and glycoproteins, which are formed in the endoplasmic reticulum and Golgi apparatus. Lignin is formed within the cell wall. Microfibril orientation is thought to be controlled by cytoplasmic microtubule orientation and perhaps also by the self-assembly properties of the wall. The main control of matrix polysaccharide synthesis is by variation of polysaccharide synthase levels. Growth substances profoundly influence both the nature and quantity of the polymers deposited in the cell wall.

References

Baydoun, E.A-H., Waldron, K.W. and Brett, C.T. (1989) The interaction of glucuronyltransferase and xylosyltransferase involved in glucuronoxylan synthesis in pea (*Pisum sativum*) epicotyls. *Biochem. J.,* **257**, 853–858.

Brown, J.M., Li, L., Okuda, K. et al. *(1994) In vitro* cellulose synthesis in plants. *Plant Physiol.,* **105**, 1–2.

Crosthwaite, S.K., MacDonald, F.M., Baydoun, E.A.-H. and Brett, C.T. (1994) Properties of a protein-linked glucuronoxylan formed in the plant Golgi apparatus. *J. Exp. Bot.,* **45**, 471–475.

Delmer, D.P., Ohana, P., Gonen, L. and Benziman, M. (1993) In vitro synthesis of cellulose in plants – still a long way to go. *Plant Physiol.,* **103**, 307–308.

Edwards, M., Scott, C., Gidley, M.J. and Reid, J.S.G. (1992) Control of mannose/glucose ratio during galactomannan formation in developing legume seeds. *Planta, 187*, 67–74.

Feingold, D.S. and Avigad, G. (1980) Sugar nucleotide transformations in plants, in *The Biochemistry of Plants,* (ed. J. Preiss), Academic Press, New York.

Ferro-Novick, S. and Jahn, R. (1994) Vesicle fusion from yeast to man. *Nature, 370*, 191–193.

Giddings, T.H. and Staehelin, L.A. (1991) Microtubule-mediated microfibril deposition: a reexamination of the hypothesis, in *The Cytoskeletal Basis of Plant Growth and Form,* (ed. C. Lloyd), Academic Press, New York, pp. 85–99.

Hobbs, M.C., Delarge, M.H.T. and Brett, C.T. (1991) Differential distribution of a glucuronyltransferase, involved in glucuronyl xylan synthesis, within the Golgi apparatus of pea (*Pisum sativum* var. Alaska). Biochem. J., *277*, 653–658.

Jarvis, M.C. (1992) Self-assembly of plant cell walls. *Plant, Cell & Env., 15*, 1–5.

Kieliszewski, M.J. and Lamport, D.T.A. (1994) Extensin: repetitive motifs, functional sites, post-transitional codes and phylogeny. *Plant J., 5*, 157–172.

Lynch, M.A. and Staehelin, L.A. (1992) Domain-specific and cell-type-specific localisation of two types of cell wall matrix polysaccharides in clover root tips. *J. Cell Biol., 118*, 467–479.

Maclachlan, G., Levy, B. and Farkas, V. (1992) Acceptor requirements for GDP-fucose: xyloglucan 1,2-L fucosyltransferase activity solubilised from pea epicotyl membranes. *Arch. Biochem. Biophys., 294*, 200–205.

Piro, G., Zuppa, A., Dalessandro, G. and Northcote, D.H. (1993) Glucomannan synthesis in pea epicotyl – the mannose and glucose transferases. *Planta, 190*, 206– 220.

Reis, D., Vian, B., Chanzy, H. and Roland, J.C. (1991) Liquid-crystal-type assembly of native cellulose-glucuronoxylan extracted from plant cell walls. *Biol. of the Cell, 73*, 173–178.

Rodgers, M.W. and Bolwell, G.P. (1992) Partial purification of Golgi-bound arabinosyltransferases and two forms of xylosyltransferase from French bean (*Phaseolus vulgaris* L.). *Biochem. J., 288*, 817–822.

Roelofsen, P.A. and Houwink, A.L. (1953) Architecture and growth of the primary cell wall in some plant hairs and in the *Phycomyces* sporangiophore. *Acta Bot. Neerland., 2*, 218–225.

Schupmann, H., Bacic, A. A. and Read, S.M. (1994) UDPGlc metabolism and callose synthesis in cultured pollen tubes of *Nicotiana alata* Link et Otto. *Plant Physiol., 105*, 659–670.

Seagull, R.W. and Falconer, M.M. (1991) *In vitro* xylogenesis, *in The Cytoskeletal Basis of Plant Growth and Form,* (ed. C. Lloyd), Academic Press, New York, pp. 183–194.

Shibaoka, H. (1994) Plant-hormone induced changes in the orientation of cortical microtubules. *Ann. Rev. Plant Physio. Mol. Biol.,* **45**, 527–544.

Waldron, K.W. and Brett, C.T. (1990) Effects of extreme acceleration on the germination, growth and cell wall composition of pea epicotyls. *J. Exp. Bot.,* **41**, 71–77.

White, A.R., Xin, Y. and Pezeschk, V. (1993) Xyloglucan glucosyltransferase in Golgi membranes from *Pisum sativum* (pea). *Biochem. J.,* **294**, 231–238.

Zhang, G.F. and Staehelin, L.A. (1992) Functional compartmentalisation of the Golgi apparatus of plant cells – immunocytochemical analyses of high-pressure freeze-substituted sycamore maple suspension-cultured cells. *Plant Physiol.,* **99**, 1070–1083.

Further reading

Bolwell, G.P. (1988) Synthesis of cell wall components – aspects of control. *Phytochem.,* **27**, 1235–1253.

Bolwell, G.P. (1993) Dynamic aspects of the plant extracellular matrix. *Int. Rev. Cytol.,* **146**, 261–324.

Delmer, D.P. (1987) Cellulose biosynthesis. *Ann. Rev. Plant Physiol.,* **38**, 259–290.

Delmer, D.P. and Stone, B.A. (1988) Biosynthesis of plant cell walls, in *The Biochemistry of Plants,* Vol. 14 (ed. J. Preiss), Academic Press, New York, pp. 373–420.

Drouich, A., Faye, L. and Staehelin, L.A. (1993) The plant Golgi apparatus: a factory for complex polysaccharides and glycoproteins. *Trends in Biochem. Sci.,* **18**, 210–214.

Lloyd, C.W. (ed.) (1991) *The Cytoskeletal Basis of Plant Growth and Form,* Academic Press, New York.

Waldron, K.W. and Brett, C.T. (1985) Interactions of enzymes involved in cell-wall heteropolysaccharide biosynthesis, in *Biochemistry of Plant Cell Walls* (eds C.T. Brett and J.R. Hillman), CUP, Cambridge, pp. 79–97.

5 The cell wall and control of cell growth

5.1 Relationship between wall properties and turgor in cell extension

The growth of plant cells depends on the interaction between the turgor pressure, which presses the protoplast outwards against the wall, and the mechanical strength of the wall, which tends to prevent any rapid increase in volume (Figure 5.1). In multicellular organisms, the pressure exerted by neighbouring cells may be important, but in the tissue as a whole the interaction between turgor pressure and wall strength controls growth. In most non-growing, primary-walled plant cells, the outward pressure produced by turgor is approximately equalled by the inward force exerted by the wall. As the cell absorbs water, the wall stress increases. The elastic energy stored in the strained bonds of the polymers, and possibly in their increased order, compresses the protoplast, resulting in turgor. As turgor increases, the wall stretches only slightly before the additional forces set up within it serve to counterbalance the increased turgor and to prevent rapid changes in volume. Only in a few cases – such as stored growth (section 5.3) and in the thigmonastic response of *Mimosa* – do rapid changes in volume occur.

However, this elastic aspect of cell walls, which keeps the cell at approximately constant volume over a range of turgor pressures and influences the turgidity of the tissue (e.g. turgid and wilted leaf tissues), is only part of the story. In growing cells, once a certain critical turgor pressure has been exceeded the wall begins to stretch in a **plastic** (i.e. non-reversible or viscous) manner. This relatively slow process is superimposed on the **elastic** (i.e. reversible) stretching of the wall; it results in a permanent increase in cell volume and hence in cell growth. The speed at which this plastic extension occurs and the parts of the wall in which it occurs are closely controlled by the plant, permitting coordinated growth of the various tissues.

One of the earliest but most useful models of plant growth was developed by Lockhart (1965). In this model (equation 5.1), the critical turgor

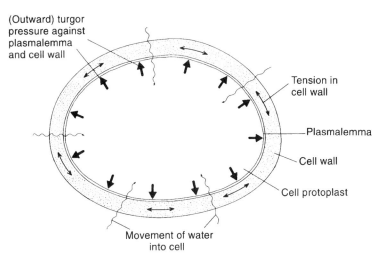

Figure 5.1   Turgor pressure on the cell wall. The influx of water down an osmotic gradient into the cell results in a positive internal pressure that presses the protoplast against the cell wall. This leads to a build up of tension (stress) within the cell wall and an increase in turgidity. Such turgor pressure is the driving force behind the extension of growing cells.

pressure above which plastic wall extension and growth occurs is known as the **threshold turgor pressure** (P_{tb}). In higher plant cells it appears to be fairly constant, not changing in response to change in turgor pressure or to plant growth regulators. The rate of cell growth is proportional to the difference between the **turgor pressure** (P_t) and the threshold turgor pressure (equation 5.1). The **constant of proportionality** (m), also known as **extensibility**, does, in higher plants, change in response to certain growth regulators and probably represents the main means of control which the plant exerts on the rate of growth of individual cells. The other major variable, turgor pressure, varies according to the availability of water. The process of water uptake may be incorporated into the model by including the **osmotic pressure difference** across the plasma membrane ($d\pi = \pi_i - \pi_o$) and the ease of water movement into the cell (L = hydraulic conductance) as depicted in equation 5.2. If $L > m$, then equation 5.2 tends towards equation 5.1. This is generally the case and has been verified experimentally (see Cosgrove, 1993). Hence, whilst water availability is important in the maintenance of turgor, the hydraulic conductance of water through a plant tissue does not appear to be an important factor in the regulation of cell growth.

Equation 5.1

$$\frac{dv}{dt} = m(P_t - P_{tb})$$

where v = cell volume, m = wall extensibility, P_t = turgor pressure, P_{tb} = threshold turgor pressure and t = time.

Equation 5.2

$$\frac{dv}{dt} = \frac{mL}{m+L}(d\pi - P_{tb})$$

where v = cell volume, m = wall extensibility, P_{tb} = threshold turgor pressure, $d\pi$ = osmotic pressure difference across the plasma membrane, L = hydraulic conductance and t = time.

Thus the main site of control of growth resides in the wall itself, at least in the short term. The fact that certain growth regulators can alter growth rate within a few minutes implies that they have a fairly rapid effect on the extensibility of the wall, though long-term effects involve changes in the rate of wall synthesis as well. To understand how these effects occur, one has to investigate the physical properties of the cell wall and how they change during a physiological process. Such investigations have involved the study of cell walls *in vitro* after isolation and *in vivo* in living tissues, using a variety of methods.

The cell wall is polymeric in nature (Chapter 2) and therefore exhibits both viscous and retarded-elastic deformations in response to stress, i.e. it is viscoelastic. Accordingly, it is logical to suppose that wall extension will, to a large extent, comprise a viscoelastic slippage of wall polymers. Such properties can be studied using isolated cell walls which have been suitably treated to eliminate any biochemically mediated processes. However if, as seems likely (section 5.5), the initiation of extension involves the biochemical (enzymatic) cleavage of load-bearing cross-links between wall polymers, such chemorheological extension could not be studied in isolated cell walls unless the enzymes concerned were reintroduced. Furthermore, the properties of an isolated wall can only reveal the effects of biochemical alterations that have occurred prior to isolation. A more realistic picture of cell wall extension can thus be obtained from the study of living tissues.

5.2 The physical properties of the isolated cell wall under tension

The physical properties of isolated cell walls are generally investigated *in vitro* by excising a whole cell, tissue or piece of cell wall, inhibiting wall enzymes with a suitable treatment, and then applying a stretching force, or stress, to it. Inactivation of enzymes may be achieved by boiling the tissue in methanol, followed by a rehydration step. While it is important to eliminate the undefined activities of enzymes within an isolated specimen, this methodology is not ideal, for cell wall dehydration can lead to new and often very strong non-covalent interactions forming within the wall, thus changing some of its physical properties. Furthermore, the boiling of tissues in methanol may introduce other chemical changes to the cell wall polymers.

The type of stress that can be applied is often determined by the tissue under investigation. In the case of higher plants and their tissues, one is limited to clamping the tissue at opposite ends and imposing a one-dimensional (**uniaxial**) stress between the points of attachment. This approach can yield much useful information on the viscoelastic properties of the cell wall. However, it cannot mimic turgor pressure which, *in vivo*, imposes an outward pressure on all parts of the wall, leading to two-dimensional (**multiaxial**) 'stresses' within the wall. Moreover, work with higher plant tissues faces problems arising from their multicellularity and cell heterogeneity.

Such problems have, to a certain extent, been overcome in experiments involving the giant cells of the characean algae. These cells are so large (Figure 5.2) that they may be excised, emptied of cytoplasm and, after connection to a capillary tube, pumped full of mercury. By regulating the pressure, a controlled multiaxial stress can be imposed upon the cell walls, which is similar to that induced by turgor pressure. This approach also overcomes problems such as 'necking' that are so often created by uniaxial stress (section 5.2.1).

Whichever method is used, the application of stress results in a characteristic extension known as '**strain**'. Three different types of stress–strain relationship have been commonly studied: measurement of strain during a period of constant stress; measurement of stress while under a constant rate of increase in strain; and stress relaxation measurements.

5.2.1 Measurement of strain (extension) during a period of constant stress

Application of constant stress gives rise to both plastic (irreversible) and elastic (reversible) deformation (Figure 5.3); the isolated wall has the properties of a viscoelastic plastic material. Initially, the wall extends as a result of instantaneous elastic deformation (De_1). This is followed by a slow yielding ('creep') which is roughly proportional to log time, and is composed of a retarded elastic component (De_2) and a plastic component (Dp). Upon removal of the load there is a partial reversal of the extension, comprising an instantaneous contraction equal to De_1, followed by a much slower one equivalent to De_2. The distance Dp can

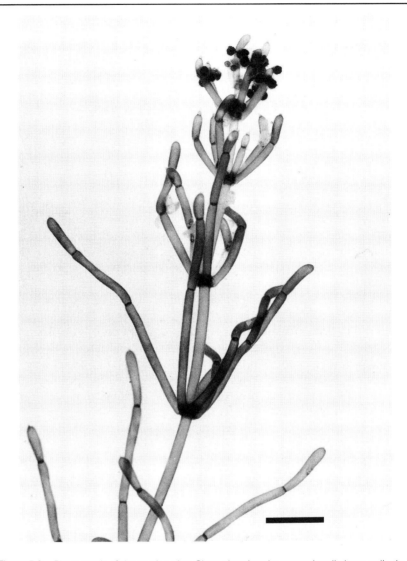

Figure 5.2   Photograph of the marine alga *Chara* showing the exceptionally large cells that have allowed the study of extension by uniaxial and multiaxial stress *in vitro*. Bar, 5 mm.

then be measured once elastic contraction has finished. From Figures 5.4 and 5.5, it can be seen that uniaxial stress results in a much greater value for Dp than multiaxial stress. This is a result of unusual structural modifications within the wall, and exaggerated realignment of cell wall components (especially microfibrils) in the direction of the uniaxial stress, a problem not encountered under conditions of multiaxial stress.

Furthermore, uniaxial stress induces 'necking' as a result of contractile forces acting at right angles to the direction of stress, which often leads to a reduction in diameter of the cell – an effect which is not characteristic of extending tissues (Figure 5.6). In such investigations, it is important to remove residual water from the sample prior to the assay. Otherwise, fluid movement out of the tissue may create a significant error in the value of Dp. Removal of excess fluid from e.g. a freeze–thawed or re-hydrated specimen may be achieved by pressing under a constant weight for a short period.

As noted previously, it is the plastic deformation (Dp) which is important for growth, since elastic extension could only result in growth under conditions of steadily increasing turgor. However, in the isolated wall, we have seen that the rate of plastic extension, under constant stress (i.e. creep) rapidly falls with time (Figure 5.5). This phenomenon, which is well accounted for by physical 'strain hardening' properties, is in marked contrast to the rate of growth *in vivo*, which remains constant for a considerable period if turgor is maintained. It is clear, therefore, that the growth process *in vivo* is not simply a passive stretching of an inert wall, but must involve some activity of the living cell in order to maintain the continued growth rate. Generally, the initial rates of creep of cell wall preparations (both of freshly isolated walls and of those that have involved a drying step) vary approximately in accordance with differences in the growth rates of the parent tissues. In view of this, extension *in vivo* may be thought of as a metabolically sustained creep process.

5.2.2 Measurement of stress while under a constant rate of increase in strain

This method, which requires the use of a machine known as an Instron analyser (Figure 5.7), has concentrated upon uniaxial stress–strain relationships. By plotting stress–strain graphs, curves like those which are illustrated in Figure 5.8 are produced. From the slopes of these curves, one can calculate the plastic compliance (DP) and elastic compliance (DE) which correspond to the per cent extension per unit stress. Generally, variations in the DP of cell wall preparations from various tissues correlate quite well with variations in the growth rates of the tissues. Changes in DP as a result of application of growth regulators (see below) also correlate well with changes in growth rate. However, the correspondence between changes in growth rate and changes in DP is not exact, so DP does not fully represent the ability of a wall to extend.

5.2.3 Stress relaxation measurement

Here, the wall is rapidly stretched to a new length and held thus while

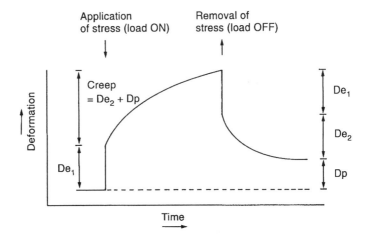

Figure 5.3   A schematic diagram showing typical in vitro cell-wall extension under a constant load. At 'load on', the wall undergoes an instantaneous elastic deformation, De_1. This is followed by a phase known as 'creep' which consists of a plastic deformation, Dp, and a retarded elastic deformation, De_2. At 'load off', the cell wall contracts to a length equivalent to its initial length plus the degree of plastic deformation.

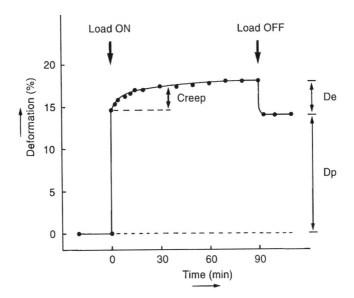

Figure 5.4   Longitudinal deformation response of a *Nitella* wall segment under uniaxial stress. The segment, isolated from a 21-mm control cell, was subjected to a load of 4.4g (equivalent to the calculated *in vivo* longitudinal force). De, elastic (reversible) deformation; Dp, plastic (irreversible) deformation.

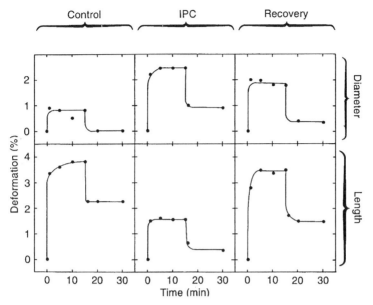

Figure 5.5   Typical deformation responses of mercury-filled *Nitella* wall tubes to multiaxial stress. An internal pressure of 5 bars was applied at 0 min and lowered to 0.5 bars at 15 min. The general features of the responses are similar to those seen in Figure 5.4. However, the per cent deformation in the length is much reduced even though the forces are equivalent. Here, walls of different structural characteristics were obtained from control tissues, isopropyl *N*-phenylcarbamate (IPC)-treated (60 h) and IPC recovery cells (IPC-treated cells that had been removed from IPC-containing media and incubated in fresh media for at least 20 h). IPC is an antimicrotubule drug, and when applied to *Nitella* elicits a shift in microfibril deposition from a generally transverse pattern to a random pattern. As a result, deformation through multiaxial stress is increased in the transverse direction, and decreased in the longitudinal direction – typically illustrating the 'multiaxial growth' hypothesis.

the decay in stress is measured (Figure 5.9). The stress decreases at a rate which can be represented by the empirical equation 3.

Equation 3

$$s = b \log \frac{t + T_m}{t + T_o} + C$$

where s = stress, t = time and b, T_m, T_o and C are constants. The constant T_o, known as the minimum stress relaxation time, is different in tissues with different growth rates and shows an approximate inverse correlation with the rate of growth when measured before and after the application of growth-promoting substances such as auxin. T_o is thought

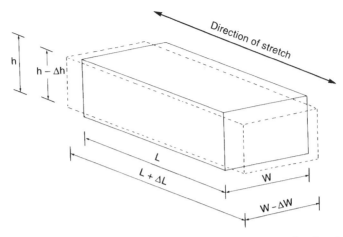

Figure 5.6 Poisson's Ratio. If one stretches a block of material in one direction, then if its volume is to remain constant, it contracts at right angles to the stretch. The contraction in width is proportional to the initial width and the increase in length such that:

$$\frac{\Delta W}{W} = \frac{\Delta b}{b} = \rho . \frac{-\Delta L}{1}$$

where ρ = constant of proportionality, and is known as Poisson's Ratio. Thus, when uniaxial stress is applied to plant tissues, the cells will contract in width as their length increases. This 'necking' results in unusual and often extreme rearrangement of many cell wall components (e.g. the alignment of cellulose microfibrils) due to the abnormal direction of forces imposed upon them. Hence, while uniaxial stress-type investigations can yield much useful information, they cannot mimic the true effects of turgor pressure.

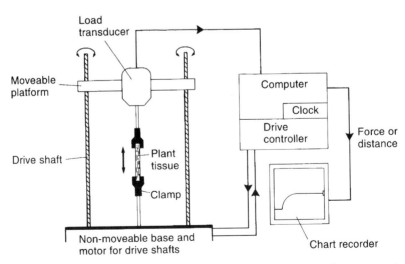

Figure 5.7 Diagram illustrating the use of an Instron analyser in measuring stress and strain in a piece of plant tissue.

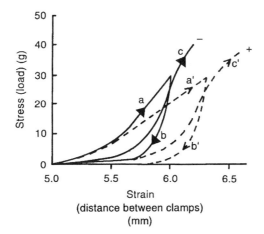

Figure 5.8   Instron extension experiment on oat coleoptile walls that have been obtained from control (–) or auxin-treated (+) coleoptiles. Solid lines: load extension curves from control tissue. Line (a) shows the change in stress whilst increasing strain at a constant rate to a final loading of 30 g. On reducing load (b) the extension decreased, but along a different path from (a). Upon re-extension to the same loading (c), it can be seen that a lower stress is required to induce a given strain when compared with (a). This is because in (a), the extension consists of an irreversible plastic component as well as a reversible elastic one. Therefore, upon re-extension (c) to the same loading, only the elastic component remains. DT, a measure of the total extensibility of the wall, is calculated from the gradient of the linear portion of initial extension curves (a). DE, the elastic compliance, is calculated from the gradient of the linear portion of the re-extension curves (c). DP, the plastic compliance, is the difference between DT and DP. Broken lines: load extension curves from auxin-treated tissue. Curves (a'), (b') and (c') are equivalent (in terms of load-extension treatment) to (a), (b) and (c), respectively. Auxin treatment of plant tissue reduces the resistance to extension from both the elastic and plastic components resulting in a shift of all curves, and increased values in DT, DE and DP. The increase in DP shows that auxin has a much larger effect on the irreversible plastic properties of the cell wall than on the elastic ones.

to be related to the molecular weight of a polymer or polymers involved in controlling the flow rate of material within the wall during wall extension. T_m is the maximum relaxation time, b is thought to be proportional to the number of viscoelastic flow units per unit cross-sectional area of the wall, and C represents the number of cross-links between adjacent wall molecules. Stress relaxation measurements yield information on the activation energies involved in overcoming the resistance of bonds which normally prevent wall extension. Such studies have, for instance, led to the suggestion that resistance to stress of *Nitella* walls is due to the cooperative actions of many non-covalent interactions rather than a relatively small number of covalent bonds.

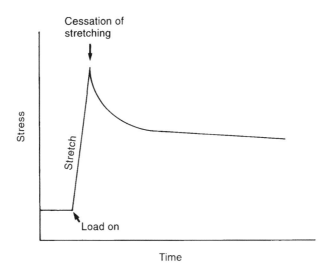

Figure 5.9 Stress relaxation of oat coleoptile cell walls. The wall is stretched to a given length, and the tension (stress) within the wall measured over a time period. As shown, after stretching has ceased, the tension within the wall decreases even though the length is being maintained at a new high value.

5.3 Physical properties of the cell wall during extension in vivo

As mentioned in section 5.1, the key difference between the behaviour of isolated cell walls under imposed stress and that of growing tissues is that extension of isolated cell walls decreases rapidly with time, whereas living cells exhibit a constant growth rate. This has led to the suggestion that a metabolic process is needed to sustain growth. This conclusion is supported by the observation that the temperature coefficient (Q_{10}) for growth is approximately 10, typical of an enzyme-mediated process, while that for extension of isolated cell walls is near to 1, characteristic of a purely physical process. Furthermore, if respiration is blocked by application of inhibitors such as azide, growth is inhibited even if turgor is not affected, again indicating that energy-requiring steps are needed for cell wall extension and cell growth.

There are several ways in which the wall yielding properties of living tissues have been investigated. The earliest approach involved measuring the growth rate of excised tissues after modifying turgor pressure, for example with the use of osmotica. Later, methods were developed to measure the growth response of growing tissues to applied mechanical (usually tensile) forces. A third approach has involved the study of stress relaxation *in vivo*.

5.3.1 Effect of turgor on growth rate

Early experiments concentrated on measuring growth rate of excised tissues after modifying turgor pressure. This was often achieved by incubating tissues in osmotica such as mannitol or polyethylene glycol and plotting growth either against cell turgor or against osmotic pressure/water potential of the incubating solution. Turgor could also be modified by allowing plants to dry down (wilt). More recently, pressure chambers have been used to modify the turgor pressure, either by enclosing the roots and using pressure to modify xylem pressure, or by enclosing the entire growing tissue. The latter methods are advantageous compared with the former, since they use intact tissues rather than relying on the ability of tissues to continue to grow and exhibit growth responses after excision.

The results of such experiments have usually been interpreted in the light of the biophysical model of plant growth discussed above (equations 5.1 and 5.2). Tissues which behave according to the model would be expected to demonstrate a linear fall in growth rate as turgor pressure is reduced (Figure 5.10). Accordingly, the slope of the line would give a value for the wall yielding coefficient (extensibility), m, and the x-intercept would give a value for P_{tb}. Experimentation has demonstrated that the growth curve of many tissues is linear with turgor. However, in some cases the curve may become steeper with increased turgor, indicating that the value for extensibility (m) is not always constant. Indeed, it has been suggested that the stimulation of extension by auxin is brought about by a change in wall extension from a linear function of turgor to non-linear (Cleland, 1959; Lockhart, 1965). In other tissues, high turgor results in a flattening of the curve, i.e. m becomes zero. It is likely that under these circumstances the limiting factor is no longer turgor but another component, e.g. polymer shear rate. A consequence of this property is that extension under these circumstances cannot be limited by hydraulic conductance. The value for P_{tb} is also unclear in some tissues as a result of a gradual increase in m at low turgor (Figure 5.10). Such problems arise from the increasing significance of wall elasticity at this portion of the curve. Separation of elastic from plastic changes often assists in the measurement of P_{tb}.

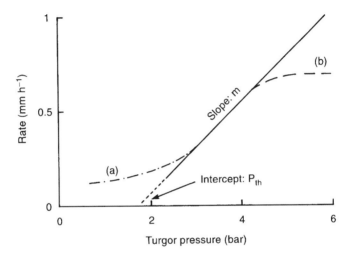

Figure 5.10 The effect of turgor on growth rate. (——) = tissues which behave according to equations 5.1 and 5.2; (·····) = tissues in which extensibility (*m*) increases with turgor; (- - -) = tissues in which extensibility (*m*) levels off at high turgor. Values for growth rate and turgor are hypothetical and represent the likely order of magnitude in immature stem tissues.

5.3.2 (Rapid) growth response of growing tissues to applied mechanical forces

This approach involves the application of tensile forces to living tissues followed by the measurement of extension immediately ensuing. The ratio of the extension response to the tension applied gives an indication of wall yielding. Essentially, the types of extension curve obtained (Kutschera and Schopfer, 1986) are similar in shape to that used to illustrate strain during a period of constant stress (Figure 5.3). Workers have used this approach to estimate the plastic (Dp) and elastic (De) components of extension and have attempted to relate data from such studies to equations 5.1 and 5.2. However, difficulties have been encountered, principally because this methodology usually measures the rapid response immediately following an externally applied force rather than the normal turgor-driven extension. Furthermore, the force will generally result in a unilateral rather than multilateral stress, as discussed in section 5.2.1.

5.3.3 Stress relaxation *in vivo*

This approach investigates the decrease in wall stress whilst preventing water uptake and cell expansion. Under these conditions, any changes in wall stress (which provides the counteracting force to turgor pressure) may be detected by a concomitant decrease in turgor. Several methods have been used to detect such changes. The most simple have involved the use of a pressure probe inserted into an excised tissue which has been sealed to prevent evaporation. In this situation, the expected relaxation of the cell wall would result in an exponential decay in turgor pressure which would eventually stop at P_{th} (Figure 5.11). This has been demonstrated experimentally in segments of pea stem tissue (Cosgrove, 1985). Interestingly, auxin induced a decrease of P_t to the same value of P_{th}, but at a faster rate (depicted schematically in Figure 5.11). More complex methods have made use of excised tissues in oil-filled pressure vessels. Recently, the development of a 'pressure block' technique (Cosgrove, 1987) has enabled such relaxation experiments to be carried out on intact tissues (Figure 5.12). In these studies, the externally applied pressure required to prevent tissue enlargement (as determined with a position transducer) is monitored. Interestingly, comparing the results from intact (pressure block method) and excised tissues indicates that, in both cases, relaxations are broadly similar, except that those of intact tissues are apparently faster and larger. In both cases, results have validated the growth model in equations 5.1 and 5.2. However, it is now apparent that the parameters m and P_{th} reflect complex underlying processes and are not simple time-invariant constants.

As stated in section 5.1, P_{th} in higher plants is of a non-variable type, and lowering P_t results in a lowering of growth rate until P_{th} is reached, at which point growth becomes zero. This is not the case in all plants. In the giant alga *Nitella*, for example, many experiments have also been carried out in which three of the four unknown parameters have been measured experimentally. Growth rate has been assessed by monitoring the distance between resin-bead markers by time-lapse cinemicrography. Turgor pressure has been measured directly by inserting a micro-manometer into the vacuole, and the threshold turgor pressure has been estimated by lowering the turgor pressure to a value at which growth ceases, at which point $P_t = P_{th}$. Modification of P_t has been achieved by varying the concentration of an osmoticum in the external medium, or by drawing off a small volume of vacuolar contents with a microcapillary tube (sap dilution method). Experiments in which P_t is

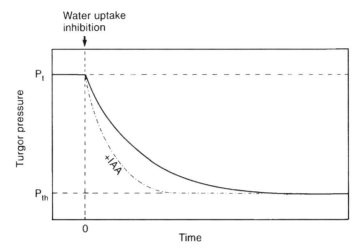

Figure 5.11 Changes in turgor pressure during in vivo stress relaxation (——). Inhibition of water uptake results in a decrease in turgor due to continued wall relaxation. If wall yielding follows equation 5.2, turgor will decay exponentially to P_{th}. The rate of decay will depend on the wall extensibility (m) and the volumetric elastic modulus of the cell. Addition of auxin (e.g. IAA) alters the rate of decay but not the value of P_{th} (-·-·-·). For further information, see text.

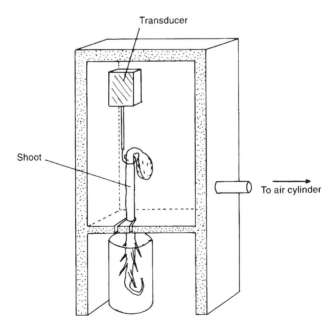

Figure 5.12 Pressure-block apparatus. The elongating region of the stem is sealed into the apparatus with a suitable epoxy adhesive and attached to a position transducer to monitor length. During wall relaxation, the length of the stem is prevented from increasing by applying the minimum pressure required (e.g. from an air cylinder).

lowered by altering the osmolality of the external medium have shown that a drop in P_t of as little as 0.2 atmospheres (from 5 atmospheres to 4.8 atmospheres) will result in an immediate cessation of growth. However, after a time lag, the length of which depends on the magnitude of the decrease, growth will resume provided that the new P_t is greater that 2 atmospheres. By examining the equation, one can see that the only way that the cell can resume extension is by lowering the value of P_{th}. Thus, in *Nitella*, one aspect of growth-rate control involves the modification of the P_{th}. Because this modification is inhibited by azide, it is presumably metabolism-dependent. If P_t is increased, growth-rate regulation also occurs, again due to alteration of P_{th}. If P_t is altered by the sap dilution method, neither P_{th} nor m change. In this situation, the P_t is raised to its original value by an osmoregulatory system, showing yet another pathway by which the cell can regulate steady state extension.

In some plant tissues, even though no growth occurs when $P_t = P_{th}$, wall-loosening appears to continue, since a subsequent rise in P_t leads to a burst of high growth rate followed by a fall to a growth rate comparable with controls. This phenomenon of 'stored growth' can usually be observed in slow-growing tissues but not always in fast-growing ones. It suggests that, at least in slow-growing tissues, the events which weaken the wall and increase its extensibility can occur independently of wall extension.

Generally, the results from studies on intact tissues have shown that wall relaxation initiates cell extension. The wall stress becomes reduced, causing the turgor pressure, and therefore the water potential, to decrease. Water is drawn into the cell, resulting in cell expansion. This can be summarized in a general model for cell extension (Figure 5.13) in which steady state growth is brought about by a balance between metabolic cell wall loosening and strain hardening processes, which retard extension. In this respect, the mechanisms (biochemical or otherwise) that alter the physical properties of the cell wall and bring about wall weakening are of great interest. This aspect is discussed further in section 5.5.

5.4 Control of wall extensibility

5.4.1 Effects of growth substances

A plant tissue or organ experiences the effects of a range of growth substances *in vivo*. Some of these growth substances are produced by the tissue itself, whilst others are transported into the tissue from other parts of the plant. Thus the removal (excision) of an organ or a piece of tissue from a plant alters, and generally decreases, the concentration of

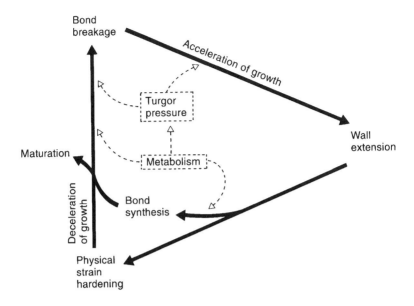

Figure 5.13   Diagram of a model for cell-wall extension. Each turn of the cycle represents a microscopic, independent viscoelastic-plastic extension step. Metabolism plays a central role, maintaining turgor pressure, mediating wall loosening as well as causing wall stiffening through bond synthesis, ultimately leading to cell maturation. In this model, turgor pressure is required for both bond lysis and for driving extension. Each cycle includes a deceleration factor of physical or metabolic origin or both.

growth substances within it. A variety of growth substances, notably auxins, gibberellins and ethylene, affect the rate of growth of organs such as stems and coleoptiles, and excision of these growth substance producing regions of these organs causes a change (generally a marked fall) in the growth rate of the remaining tissues.

Auxin has been studied most closely with respect to its effect on growth rates. The organs used have commonly been coleoptiles (e.g. of oat), epicotyls (e.g. of pea) and hypocotyls (e.g. of mung bean). The effect of auxin on such excised organs is to cause a dramatic increase in growth rate after a lag period of between 10 and 20 minutes, with a maximum rate being attained after about 30 minutes (Figure 5.14). The initial high rate is usually transient and soon falls to a constant rate that is at a high level relative to the controls. The rate may then decline gradually over a period of up to 24 hours. Experiments on rye coleoptiles of the sort described in section 5.3 have shown that this stimulation of growth by auxin is mediated not by a change in the threshold turgor pressure (P_{th}), but by an increase in the value of m, i.e. the extensibility of the cell walls.

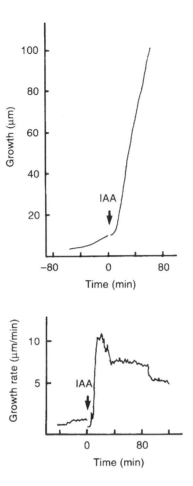

Figure 5.14   The effect of auxin on the growth of an oat coleoptile segment. (a) A typical growth curve of a coleoptile segment, showing the effect of IAA at 10^{-5}M on the increase in length. (b) As in (a), except that growth rate instead of growth is plotted against time.

The effect of auxin may be complicated by a process of feedback inhibition involving an oligosaccharin. The nonasaccharide that forms a subunit of many xyloglucans has been found to inhibit the auxin-induced growth of pea epicotyls. This nonasaccharide, which contains four residues of glucose, three of xylose, and one each of galactose and fucose (Figure 2.16 and Box 5.1), is inhibitory at very low concentrations, around 10–100 ng ml^{-1}. It is produced by the action of an endoglucanase on xyloglucan, and the activity of this endoglucanase in pea epicotyls rises in response to auxin. Hence the nonasaccharide is produced as an indirect effect of auxin action and acts to inhibit, at least

Box 5.1 Xyloglucan oligosaccharins: structure and nomenclature

Xyloglucans are important primary wall components (sections 2.6.5 and 3.3.1) and act as food reserves in seeds (section 8.5.2). Oligosaccharides derived from them have a range of effects as cell-signalling molecules – i.e. they act as 'oligosaccharins' (Box 5.2). Like the parent molecule, the xyloglucan oligosaccharins have a backbone of β-1,4-linked glucose residues, most of which bear a mono-, di- or trisaccharide side-chain. Recently a systematic nomenclature has been introduced for these oligosaccharides, in which each glucose, plus its side-chain, is given a symbol according to the nature of its side-chain (Fry *et al.*, 1993b). The main symbols are given below, together with mnemonics which are intended to help in memorizing them.

Code letter	Structure represented	Mnemonic
G	β-D-Glc*p**	Glucose
X	α-D-Xyl*p*-(1–6)-β-D-Glc*p**	Xylose
L	β-D-Gal*p*-(1–2)-α-D-Xyl*p*-(1–6)-β-D-Glc*p**	gaLactose
F	α-L-Fuc*p*-(1–2)-β-D-Gal*p*-(1–2)-α-D-Xyl*p*-(1–6)-β-D-Glc*p**	Fucose
A	α-L-Ara*f*-(1–2)-β-D-Glc*p** 6 ↑ α-D-Xyl*p*-1	Arabinose
B	β-D-Xyl*p*-(1–2)-β-D-Glc*p** 6 ↑ α-D-Xyl*p*-1	Beta-xylose
C	α-L-Ara*f*-(1–3)-β-D-Xyl*p*-(1–2)-β-D-Glc*p** 6 ↑ α-D-Xyl*p*-1	follows A and B
S	α-L-Ara*f*-(1–2)-α-D-Xyl*p*-(1–6)-β-D-Glc*p**	Solanaceae

*In each case, the same symbol is used whether the glucose residue is internal or at the reducing terminal.

Examples of some known xyloglucan oligosaccharins are as follows (see the above definitions for full details of the linkages):

```
Glc–Glc–Glc–Glc      Glc–Glc–Glc–Glc      Glc–Glc–Glc      Glc–Glc–Glucitol
 ↑   ↑   ↑            ↑   ↑   ↑            ↑   ↑            ↑   ↑
Xyl Xyl Xyl          Xyl Xyl Xyl          Xyl Xyl          Xyl Xyl
         ↑                    ↑
        Gal                  Gal
         ↑
        Fuc
      XXFG                 XXLG                XXG             XXGol *
```

*The suffix 'ol' denotes a sugar alcohol, i.e. the carbonyl/hemiacetal group has been reduced to a primary alcohol.

Box 5.2 Xyloglucan oligosaccharins: biological activities

1. Inhibition of auxin-induced growth

The oligosaccharin XXFG (formerly called XG9; see Box 5.1 for terminology) inhibits the growth-promoting effect of the auxin 2,4D on pea stem segments. This inhibitory action absolutely requires the presence of fucose, so XXLG and XXXG are inactive. On the other hand, FG is effective. The optimum concentration of XXFG is 10^{-9} mol dm^{-3}; at concentrations above 10^{-7} mol dm^{-3}, the growth-promoting effect is seen (see below). The inhibitory effect cannot be overcome by increasing the auxin concentration. The mechanism of inhibition is not known; the oligosaccharin may act by binding to a receptor in the plasma membrane, since it does not easily penetrate into the interior of the cell. XXFG is known to occur naturally in plant tissues in nanomolar amounts, and accumulates in spinach cell culture to around 10^{-7} mol dm^{-3}. Thus it is likely to be significant in controlling growth *in vivo*. Growth induced by acid is also inhibited; however, the oligosaccharin concentration required is 10^{-6} mol dm^{-3}, and the structural requirements are slightly different from those needed for inhibition of auxin-induced growth, since fucose is not essential (Aldington *et al.*, 1992).

2. Growth stimulation

Certain xyloglucan oligosaccharides promote the elongation of pea-stem segments in the absence of externally applied auxin. For this effect, the optimum concentration is around 10^{-6} mol dm^{-3}, so it is less likely to be significant *in vivo*. The minimum structural unit with growth-stimulatory activity is XXXG; fucose is not required. The mechanism of stimulation is thought to be by acting as an acceptor substrate for XET (section 5.5.2). The transfer of part of a xyloglucan molecule to a short-chain oligosaccharide rather than to another xyloglucan polysaccharide would cause a drastic reduction in molecular weight; this would make the xyloglucan less able to crosslink cellulose microfibrils, and hence increase wall extensibility.

3. Effects on morphogenesis

In cultured wheat embryos, the oligosaccharin FG interacts with auxin to produce different effects. In the presence of 2,4D, 10^{-8} mol dm^{-3} FG greatly increases the number of adventitious roots. In the absence of 2,4D, the same concentration of FG increases callus proliferation (Pavlova *et al.*, 1992).

partially, the more direct effect of auxin in stimulating growth (McDougall and Fry, 1989; Box 5.2). Clearly there is a complex control system in operation here, which is as yet only partially understood. It may be that the oligosaccharin is responsible for the decline in the auxin-induced growth after the initial maximum. Until recently, the action of auxin in controlling the extension of stems was thought to occur principally by altering the extensibility of epidermal cell walls. This is because the wall extensibility of the epidermal cells was found to be lower than that of the cortical cells in some plants, and hence the extensibility of the whole organ was thought to be limited by the relative inextensibility of the epidermis. The influence of epidermal extension is discussed further in section 5.6.

Gibberellic acid (GA) is also found to induce rapid growth in certain tissues. For example, in lettuce hypocotyl sections, the growth profile produced by GA shows a lag of approximately 10 minutes and then a rise in growth rate to a new steady state within an hour (Figure 5.15).

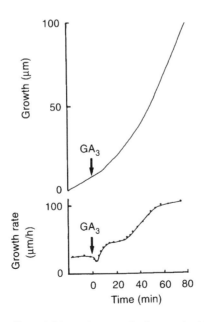

Figure 5.15   The effect of GA_3 on the growth of an excised lettuce hypocotyl. (Top) A typical growth of a hypocotyl showing the effect of GA_3 at 10^{-6} Mol dm^{-3} on the increase in length. (Bottom) As in (Top), except that growth rate instead of growth is plotted against time.

5.4.2 The acid growth hypothesis

The mechanism by which auxin can stimulate cell wall extension has been the subject of much research. Auxin has no effect on the extensibility of isolated (dead) cell walls, and hence must operate via some action on the protoplast. The long-term (>30 minutes) effects of auxin include the stimulation of synthesis of wall polysaccharides and this is probably necessary for the maintenance of high growth rate. However, the rapidity of its effect on growth rate argues for an auxin effect that is more direct than those operating via enhanced polysaccharide synthesis, which in turn probably depends on enhanced synthesis of the polysaccharide synthase enzymes.

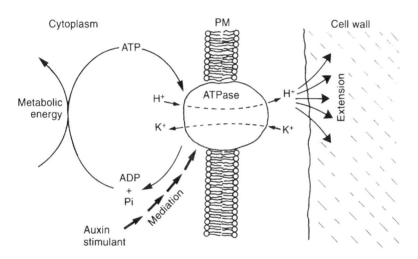

Figure 5.16   Schematic representation of a proton-pumping ATPase involved in regulating the pH of the cell wall and hence extension.

The best established theory for the rapid effects of auxin on coleoptile and stem growth is the acid growth hypothesis (Rayle and Cleland, 1992). This postulates that the effect of auxin on the cell wall is mediated by an acidification of the wall, which is due to increased activity of proton-pumping ATPase enzymes in the plasma membrane (Figure 5.16). The increased proton concentration in the wall space brings about increased wall extensibility, leading to wall extension and cell growth. Evidence for this sequence of events is summarized as follows:

(a) Auxin is known to stimulate the rate of proton extrusion across the plasma membrane and this stimulation occurs rapidly enough to be the cause of increased extensibility. The effect is not due to a direct interaction between auxin and ATPase, since no such direct effect has been observed *in vitro* and since other agents, notably fusic-cocin, do have a direct effect on the ATPase and stimulate proton efflux and associated extension with no detectable lag. Conversely, agents which inhibit the plasma membrane ATPase also inhibit the growth-promoting effects of auxin.

(b) When acid buffers are applied to auxin-sensitive tissues, growth is stimulated. The pH required is around 5.5 or less, and such pH values can be expected to be achieved by the action of a plasma membrane ATPase. The effect of such buffers can only be seen if the cuticle is abraded to allow ready access of the buffer to the cells below. The effect is seen both in living tissue sections and in frozen–thawed sections. Conversely, the application of buffers of neutral pH to auxin-treated tissues inhibits the growth-promoting actions of auxin in some plants.

(c) The giant cells of the alga, *Nitella*, provide further evidence for the theory. These cells grow only at certain points along the surface and it has been shown that these points correspond to regions of low wall pH (Figure 5.17).

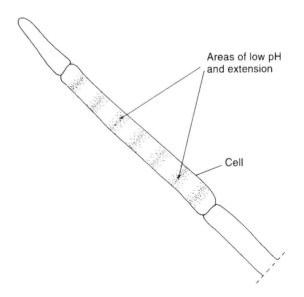

Figure 5.17 Diagram depicting the distribution of 'acid' regions and associated extension in the cell wall of *Nitella*. The regions of shading represent the areas of low pH and associated extension.

Thus the evidence for the acid growth theory is strong but extensive testing of the theory has, however, led to a number of modifications to it. First of all, the effect of acid is much shorter-lived than that of auxin (less than 2 h, as opposed to 24 h for auxin). Hence acidification of the wall must be accompanied by, or followed by, other auxin-induced effects in order for long-term growth to be achieved. Secondly, acid is not always the factor limiting growth. If, for instance, living coleoptile sections are treated with acid, they will extend rapidly for no more than 1–2 h. However, if external tension (stress) is applied, then growth may be sustained for up to 6 h. This suggests that turgor becomes limiting after a few hours of acid-induced growth. Wall extensibility may also be limited by the energy status of the tissue. Thirdly, the techniques used to demonstrate acid growth (i.e. application of acid to abraded, excised tissues) do not precisely mimic proton efflux, since proton efflux involves not only a change in wall pH but also a change in membrane potential that is important for metabolic processes such as solute transport associated with the maintenance of turgor. Hence conclusions drawn from experiments using these techniques may not be fully valid. Finally, the lack of any direct action of auxin on the plasma membrane implies that its primary action is elsewhere, and it has recently been suggested that the acidification of the wall occurs as a result of prior acidification of the cytoplasm which stimulates proton efflux.

Therefore, with these reservations and accepting that, generally, acid regulates growth only when it is the limiting factor, the acidification of the cell wall remains a well established part of our current ideas as to the mechanism of auxin action on wall extensibility. However, it must be noted that induction of rapid extension of some tissues by other plant growth substances (e.g. gibberellins, zeatin and ethylene) is not always accompanied by enhanced rates of proton extrusion, and in some cases any proton extrusion is actually inhibited during promotion of elongation. It is likely, therefore, that there are factors other than proton transport which need to be understood before we can explain extension.

5.5 Chemistry and biochemistry of cell extension

5.5.1 Changes in cell wall chemistry during cell extension

5.5.1.1 Wall turnover

Many cell wall components undergo turnover during extension growth, maturation of tissues and secondary thickening. The process usually involves the degradation of components within the wall, or removal of components from the wall by a combination of solubilization and diffusion. Turnover also occurs during terminal development such as during fruit ripening and seed germination, two aspects of which are considered fully in Chapter 9.

Investigations into the turnover of cell wall components in growing tissues have been stimulated by the interest in the underlying chemistry and biochemistry of cell wall extension. Some of the earliest evidence for turnover comes from pulse-chase studies in which radioactively labelled sugars such as glucose or sucrose were fed to tissues for a limited time, after which the tissues were incubated in a 'chase' medium containing identical but non-radioactive sugars. A decline in radioactivity in wall polysaccharides during the chase period provided evidence of turnover (Figure 5.18). Further information could be obtained by investigating the chemistry of the labelled polysaccharides (e.g. molecular weight).

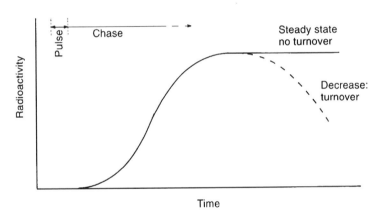

Figure 5.18 Incorporation of radioactivity into cell wall polymers during pulse-chase labelling. If the labelled polymers are in a steady state, the incorporated radioactivity will remain constant (——). However, if the polymers are turned over and solubilized from the wall, the radioactivity will decrease (– – – –).

In order to relate wall turnover to the physiological state of the cell or tissue, studies have been carried out to compare the extensibility or rate of extension of a tissue with the turnover and chemical and physical properties of constituent cell wall components. In some cases, comparisons have also been made with aspects of wall biochemistry (section 5.5.2). A common approach has involved the use of auxin to stimulate stem extension. Early studies (Labavitch and Ray, 1974) showed that in pea epicotyls fed with ^{14}C-labelled glucose, a xyloglucan fraction of the cell wall was specifically turned over during auxin-induced elongation. Further evidence for xyloglucan turnover came subsequently from studies on pea epicotyl segments which had been labelled with ^{14}C sucrose. The segments exhibited an auxin-dependent release of polymeric ^{14}C-xylose and ^{14}C-glucose in the form of xyloglucan polysaccharide. This release occurred in the presence of 0.2 mol dm^{-3} mannitol, which inhibits auxin-promoted growth, but not wall weakening (section 5.3), and not in the presence of calcium ions at concentrations that inhibit both auxin-induced extension and wall weakening. Radioactive xyloglucan was also released in response to acid at concentrations which induce extension.

In addition to influencing their turnover, auxin has also been shown to cause a decrease in the molecular weight of xyloglucans in several tissues, e.g. oat coleoptiles (Inouhe *et al.*, 1984) and azuki beans (Nishitani and Masuda, 1983) and pea epicotyls (Talbot and Ray, 1992), giving support to the view that bonds involving xyloglucans are important in the control of wall extensibility in these plants. This has been strengthened by the observation that treatment of stem segments with xyloglucan-binding lectins or antibodies (which might prevent access by a degrading enzyme) prevents auxin-induced extension (Hoson and Masuda, 1989).

In some plants, such as maize and other members of the Poaceae, similar studies have shown that β-glucan turnover and metabolism are involved in auxin-induced extension. This has been supported by studies in which antibodies to glucans and glucanases have been shown to inhibit such auxin-stimulated extension and glucan degradation (Hoson *et al.*, 1992).

Interestingly, recent work on cucumber hypocotyls undergoing natural (non-auxin stimulated) extension, indicates that there is no clear relationship between the molecular weights of xyloglucans and the mechanical properties of outer stem tissues (Wakabayashi *et al.*, 1993). This, in addition to the differences in cell wall metabolism between inner and outer tissues, indicates that the mechanical properties of some stem tissues do not depend on the molecular weight profiles of xyloglucans. The mechanism of auxin-stimulated growth may, therefore, differ from that of normal extension growth.

Auxin stimulation of xyloglucan turnover contrasts with other commonly observed instances of cell wall turnover, such as that of galactose which, although accompanying extension, does not precede it, but may be important in prolonging auxin-induced extension. Naturally occurring stem extension is generally accompanied by a decrease in the levels of pectic neutral sugars, particularly galactose.

Turnover may also be detected by analysing the changes in the composition of whole cell walls of growing tissues as they age basipetally. However, such turnover is often masked since the changes in relative levels of sugars will result not only from turnover but also from synthesis. This is demonstrated in azuki bean epicotyls in which the level of cell wall galactose remains relatively constant along the growing region of the stem. However, excision of the stem from the parent plant results in a dramatic fall in the cell wall galactose. This can be prevented by providing a sucrose source which facilitates the synthesis of new galactose-rich polysaccharides. A similar situation occurs in excised asparagus spears. Excision of the stem from the parent plant terminates the food supply. This is accompanied by a rapid decrease in the levels of cell wall galactose which would otherwise remain at a constant level in all parts of the stem.

5.5.1.2 Autolysis

The chemistry and biochemistry of cell wall turnover has been investigated in greater depth by studying wall autolysis *in vitro*. This approach involves the preparation of cell walls, free of intracellular contents, followed by their incubation in a selected buffer. During the incubation, autolysis by cell wall-degrading enzymes results in the release of carbohydrate into the medium (Figure 5.19). This is characterized for its composition, linkage and molecular weight, etc. Such studies have highlighted the turnover of pectic polysaccharides (particularly galactose and arabinose), small quantities of xylose and glucose in dicotyledonous plants including pea and chickpea stems (Seara *et al.*, 1988; Revilla *et al.*, 1986) and mixed-linkage β-glucans in monocotyledonous plants such as oats, rice, maize (e.g. Inouhe and Nevins, 1991). Detailed studies have also investigated the effects of treating the tissues with auxin prior to the autolysis experiments. This enhances autolysis of β-glucans from cell walls of maize coleoptile segments (Inouhe and Nevins, 1991) and polymers from chick pea that contain arabinose and galactose (with some xylose and glucose) (Revilla *et al.*, 1986). Further work has involved the solubilization and assay of enzymes that promote autolysis. The activity of several are increased by pre-treatment with auxin.

Whilst autolysis *in vitro* may identify the presence of a wall degrading enzyme or enzymes within the wall, it does not follow that these enzymes exhibit the same levels of activity *in vivo*.

5.5.2 Changes in cell wall biochemistry during cell extension

How, then, does acidification of the cell wall (section 5.4.2) bring about an increase in wall extensibility? It is reasonable to suppose that one or more types of bond within the wall are weakened by acid, either directly or indirectly. As yet, there is little information as to the nature of these bonds. The only part that seems reasonably certain is that the bonds con-

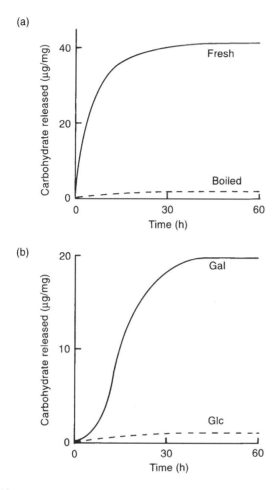

Figure 5.19   Cell wall carbohydrate release during autolytic activity in vitro. (a) Release of total carbohydrate during autolysis in freshly prepared cell walls (————) and boiled cell walls (– – –). In the latter, inactivation of cell wall enzymes inhibits autolytic activity. (b) Typical representation of cell wall sugars released from cell walls of primary-walled tissues during autolysis; pectin-derived galactose (and frequently arabinose) are often released, whilst glucose tends to remain within the cell walls.

cerned are in the matrix, or at the microfibril–matrix interface, rather than within the microfibril. If a covalent bond is involved, the most likely way for acid to influence it is via the action of an enzyme with an acid pH optimum. The cell wall contains a variety of enzymes, some of which act on cell wall components. Those investigated include glycosidases, which cleave single sugar residues from polysaccharides; glycanases, which split internal bonds within polysaccharides; and transglycosylases, which transfer polysaccharide fragments (which may contain any number of sugar residues) from one site of glycosidic attachment to another. Of these three types, the transglycosylases have, until recently, been the most attractive candidates, since they might permit wall extension without bringing about any permanent weakening of the wall (Figure 5.20). Several transglycosylases have been identified in cell walls, including a reversible α(1–6)glucanase, which also acts on arabinogalactan, and another which acts on xyloglucans, known as xyloglucan endotransglycosylase (XET). In the case of the α(1–6) glucanase, there is no evidence for an acid pH optimum; the activity of the enzyme does rise in response to auxin, but only over a period of hours, so it is not a candidate for mediating the rapid action of auxin. In contrast, examples of XET have been identified in rapidly elongating plant tissues. In several plants, XET has been shown to have an acid pH optimum of 5.5 and exhibits less than half the activity at pH 7. XET activity has been positively correlated with growth rate in different parts of the epicotyl, although the amount of extractable activity decreased after treatment with auxin (Fry *et al.*, 1992).

Studies such as those described in section 5.5.1 have highlighted auxin-stimulated turnover of xyloglucan in monocotyledonous and dicotyledonous plants and in gymnosperms. Furthermore, xyloglucan turnover after auxin treatment of, for example, pea epicotyl segments does occur rapidly enough to precede elongation (i.e. within 15 minutes). It has therefore been proposed that hydrolysis or dissociation of xyloglucan from cellulose microfibrils is the likely rate-limiting step for extension growth. Because changes in the measurable activity of cellulases and other possible enzyme candidates do not correlate with the turnover, and increases in H^+ ion concentration that induce extension would not be enough to cause release by hydrogen-bond breakage, XET has been suggested as a likely agent responsible for xyloglucan turnover and wall loosening.

Glycosidases in the wall include β-galactosidases and β-glycanases, but neither has an acid pH optimum and neither increases in activity in response to auxin. Specific inhibitors of glycosidases do not inhibit auxin-induced growth. Glycanases such as β-glucanase and xyloglucanase are present in the wall, and these probably play a role in cell wall turnover (see below). Some, e.g. cellulase, may show changes in extractibility during auxin-induced extension (Hayashi *et al.*, 1984); however, they do not appear to respond directly to acid.

5.5.2.1 Expansins

Hence, out of the above studies, XET emerges as a clear-cut favourite for an enzyme responsible for acid growth. However, non-covalent

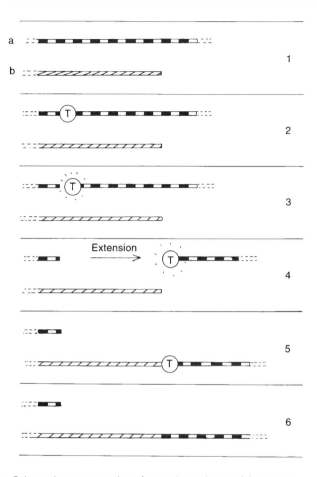

Figure 5.20   Schematic representation of transglycosylase activity on two polysaccharides (a) and (b) during extension in a hypothetical cell wall. (1) The two polysaccharides. (2) A transglycosylase becomes attached to (a). (3) The transglycosylase hydrolyses a glycosidic link in (a) and in doing so, becomes an activated intermediate while remaining attached to (a). (4) Extension separates the two pieces of (a) and the activated transglycosylase finds a suitable end of a second polysaccharide (b). (5) The activated transglycosylase catalyses the formation of a glycosidic link between (a) and (b) and then (6) becomes detached.

bonds provide further possibilities for the growth-controlling bond. Individual non-covalent bonds are relatively weak but there are many possibilities for combined action of a number of non-covalent bonds. For example, considerable numbers of hydrogen bonds are likely to be involved in binding xyloglucan and other hemicelluloses to cellulose, and the calcium bridges between polygalacturonate molecules act in a coordinated fashion. Both these types of bond have been regarded as possible candidates for growth control. Studies have shown that hydrogen-bonding of xyloglucan to cellulose appears to be unaffected by changes of pH within the physiological pH range, and agents such as urea, which disrupt hydrogen bonds, have little effect on wall extensibility. This would indicate that hydrogen bonds are not likely to be important in controlling extension. However, recent work suggests otherwise. Detailed investigations into the acid-extension response of cucumber cell walls have shown that creep in frozen–thawed stem sections is inhibited, for example, by boiling water, pre-incubation with proteases, or buffers at a neutral pH. These and other data have indicated that rugged cell wall proteins mediate the acid-extension response, probably by a chemorheological mechanism. Further work, involving a reconstitution approach, has demonstrated that a crude protein extract from the cell walls of growing regions of cucumber seedlings can induce the extension of isolated cell walls. Fractions of this extract have revealed two proteins with relative molecular masses of 29 and 30 kDa. Each can induce cell extension without detectable hydrolytic wall breakdown and can mediate the acid growth response (McQueen-Mason *et al.*, 1992). These proteins have been named **expansins**. In spite of the considerable interest in XET as a potential mediator of cell extension, these proteins do not exhibit XET activity. Furthermore, purified XET has not been shown to induce cell extension in isolated cell walls (McQueen-Mason *et al.*, 1993). Interestingly, cell wall autolysis (section 5.5.1.2) and cell wall hydrolases do not seem to play a major part in the extension of isolated cucumber cell walls. They do, however, sensitize the cell wall to expansin action (Cosgrove and Durachko, 1994).

The mode of action of expansin is not clear. It has been shown that purified expansins mechanically weaken pure cellulose paper without detectable cellulase activity. Such paper is thought to derive its strength from interfibre hydrogen bonds. This has led to the hypothesis that expansins induce cell extension by breaking hydrogen bonds between the cellulose microfibrils. This is strengthened by the observation that expansin-mediated cell extension is enhanced by 2 M urea (which would weaken hydrogen-bonding between cell wall polymers) and reduced by D_2O, which forms stronger hydrogen bonds.

Calcium bridges provide a further possibility in the control of extension since changes in pH in the physiological range cause a displacement of calcium from the wall, and in certain cases Ca^{2+} can inhibit auxin-induced extension. However, it is not certain whether the calcium ions concerned are bound to polygalacturonic acid or to some other negatively charged species, such as glycoprotein or glucuronoxylan. Another possibility for growth-controlling non-covalent bonding is provided by the presence of lectins in the cell wall. These are proteins which specifically bind certain sugars and, since many of them have

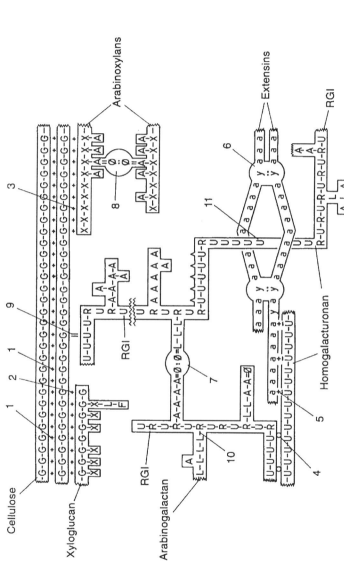

Figure 5.21 Representative primary structures and possible cross-links of wall polymers. This is not a model of the plant cell wall, and no significance is placed on the chain length, orientation, conformation, or spacing of the molecules. The diagram illustrates: (+) Hydrogen bonds: (1) cellulose–cellulose, (2) xyloglucan–cellulose, (3) xylan–cellulose, (o) calcium bridges: (4) homogalacturonan–homogalacturonan; (±) other ionic bonds: (5) extensin–pectin; (:) coupled phenols: (6) extensin–extensin, (7) pectin–pectin, (8) arabinoxylan–arabinoxylan; (=) ester bonds: (9) pectin–cellulose; (−) glycosidic bonds: (10) arabinogalactan–rhamnogalacturonan; (φ) entanglement (concatenation): (11) pectin-in-extensin. Sugars: G, glucose; X, xylose; U, galacturonic acid; A, arabinose; R, rhamnose; L, galactose; a, amino acid; y, tyrosine; y:y, isodityrosine; Ø, ferulic acid; Ø:Ø, diferulic acid.

more than one sugar binding site, they could be involved in cross-linking polysaccharides or glycoproteins. At least one such wall lectin, that of mung bean, is inhibited by acid pH. Finally, the effect of acid on the conformation of cell wall polymers *in situ* is largely unknown and major changes in such conformations could considerably affect the extensibility of the wall, either by changing the ability of the polymer to form other non-covalent bonds or simply by causing a change in the degree of entanglement of the polymers. These latter possibilities have not been investigated to any great extent. A summary of the types of bond present within the wall is given in Figure 5.21.

5.5.3 Longer-term changes in wall extensibility

The previous sections deal mainly with short-term changes in wall extensibility, particularly in response to growth substances. Over periods of hours or days, wall extensibility changes as a result of wall maturation or in response to environmental factors. These changes, such as the long-term effects of light, may also be mediated by growth substances. Clearly, growth substances have long-term as well as short-term effects on wall extensibility.

Wall extensibility generally declines as cells age. This can be seen by monitoring changes in the capacity for acid-induced wall loosening, which decreases with age but is increased by auxin. Thus auxin not only increases the acidity of the wall but also increases the ability of the wall to extend in response to acid. Calcium ions decrease tissue extensibility but it is uncertain whether this is due to direct competition with hydrogen ions or to indirect inhibition, e.g. by decreasing the activity of a wall-loosening enzyme or by enhancing interpolymeric cross-linking. It is likely that the cessation of extension is accompanied by the formation of interpolymeric cross-links which might inhibit or nullify the activity of the agent(s) promoting extension in the immature cell wall. Since the process of extension is often followed by differentiation, discrimination between events associated with differentiation and cessation of elongation may prove to be difficult. In any case, they may be intimately linked.

Most studies concerned with the cessation of extension have, as for turnover, investigated the coincidence of various chemical and biochemical events with the decrease in extensibility. As a consequence, there has been much interest in changes in phenolic cross-links between polymers as a possible point of control of wall extensibility. Two such cross-links may be involved, that between tyrosine residues within extensin (isodityrosine) and that between ferulic acid substituents in pectin (diferulic acid). Both bonds are formed by the action of peroxidase and there are several examples of an inverse correlation between wall extensibility and peroxidase content. There is also a decrease in peroxidase secretion into the cell wall in situations such as gibberellin application to spinach cells and IAA application to pea epicotyls, in which wall extensibility is increased. In the latter, it has been suggested that IAA inhibits peroxidase-dependent lignification.

In dicotyledonous plants, the amount of extensin in the cell wall generally tends to increase as growth rates decrease, making it a plausible candidate for locking the cell wall and terminating extension. Hydroxyproline-rich glycoproteins also increase in response to wounding and pathogen invasion. In monocotyledonous plants, it has been suggested that simple phenolics such as ferulic acid and related compounds are important in the termination of growth and the fixing of cell

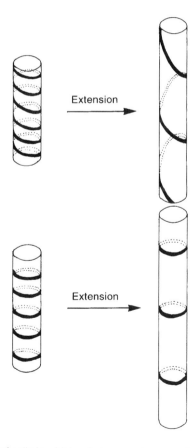

Figure 5.22   Extension of cells in which spiral or annular secondary thickening has occurred.

shape since they increase in the walls of maturing cells. However, direct causal relationships are hard to establish.

Phenolic links are also important in lignin formation, which accompanies the cessation of elongation in all cells in which lignin is laid down over the whole area of the wall. In a few cells, such as some primary tracheids and some phloem fibres, only part of the wall area is lignified and in these cases extension of the non-lignified regions of the wall can continue (Figure 5.22). The effect of lignin in inhibiting elongation is probably due both to the intrinsic strength of the numerous cross-links in lignin and also to the decrease in water content that accompanies lignification, since water probably lubricates the extension process.

Recently, the potential role of cell wall ester cross-links has also been highlighted. During elongation of maize coleoptiles, non-methyl uronyl esters between pectic rhamnogalacturonans and other as yet unidentified cell wall components increase (Carpita and Gibeaut, 1993). Accumulation of these linkages coincides with the covalent attachment of polymers containing GalA residues to the cell wall. In some cell walls, the esterification of simple phenolics from CoA-linked precursors (e.g. feruloyl-CoA) directly into the cell wall has been described. This complements the insertion of previously synthesized phenolic-substituted polysaccharides into the cell wall and would allow newly inserted phenolic monomers to serve as potential substrates for further peroxidase-catalysed ether cross-linking.

Hence, as cell extension decreases, there is an increase in interpolymeric cross-linking. Such cross-linking may reduce cell extensibility by mechanically locking the critical load-bearing polymers in place, thereby negating the effect of other mediators of extension. This may explain why, in some tissues, xyloglucan and β-glucan continue to undergo turnover after extension has ceased (see references in Carpita and Gibeaut, 1993). On the other hand, extension may cease due to a reduction in the activity of those molecules that induce cell extension.

5.6 Tissue tension

When a section of stem from the sub-apical extending zone is partially split along the longitudinal axis and then immersed in water, the stem halves curve away from each other. This effect, which occurs only in the growing regions, has led to the notion that the epidermis or epidermal region mechanically limits growth, and that it may be a target tissue for auxin stimulation. However, it is evident that the inner tissues are also able to respond to auxin, as shown by investigating extension in peeled stem sections, and the age-old observation that auxin enhances the curling of etiolated pea split-stems (which led to a commonly used bioassay for auxin). Interestingly, removal of the epidermal tissues from etiolated bean epicotyl (sub-apical) sections does not prevent curling. In these stems, the curling seems to be due to the different extension rates of the inner parenchyma tissues and outer ring of vascular tissues. Furthermore, a difference in tissue tension between inner and outer tissues is indicated by the observation that, during the cutting of the apical stem tissue in air, it immediately springs open. This presumably results from an immediate elongation of inner cells involving a change in the shape of the inner (parenchyma) cells or outer (epidermal) tissues.

5.7 Environmental influences

In seedlings, either the shoot or the hypocotyl normally extends rapidly in the dark, in order to carry the shoot apex above the soil as quickly as possible. Once the apex is exposed to light, elongation slows down (Hart, 1988). As a result of mainly morphological studies, it was agreed some time ago that the effect of light is probably comparable to the effect of ageing on cells, i.e. light accelerates cell maturation. In the same way that a range of plant growth substances including auxins have been used to investigate cell extension via the stimulation of cell elongation, light treatment has recently been employed to modulate the cessation of cell growth (Figure 5.23). Light-treated etiolated runner bean

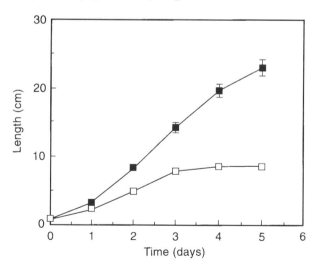

Figure 5.23   Continued growth of top 1 cm stem section of 7-day-old etiolated bean seedlings in darkness (■) or under 'grolux' daylight (□). Light treatment results in a reduction of extension rate.

seedlings demonstrate greater branching of pectic polysaccharides (on pectic backbone), and increased OH-pro-rich glycoproteins and simple phenolics, compared with their dark-grown counterparts (Waldron and Selvendran, unpublished). In etiolated rice coleoptiles, white light results in an increase in diferulic acid on arabinoxylans. In etiolated pea seedlings, white light increases the degree of lignification. Thus, the effect of light is coincident with an increase in the number of phenolic cross-links in cell walls, in both monocotyledonous and dicotyledonous plants, and a decrease in the rate of wall polysaccharide biosynthesis.

Gravistimulation may also affect the composition of cell wall phenolics in plants, as demonstrated during the growth of etiolated pea seedlings at up to 1300 x g; however, a wound response cannot be ruled out.

Summary

Cell wall extension is controlled by the degree of resistance put up by the wall against the stretching effects of turgor pressure. Various *in vitro* measurements of wall extensibility have been made, some of which are correlated with *in vivo* growth rates. Methods have been developed to investigate extensibility *in vivo*. Wall extensibility is influenced greatly by growth substances, which may exert some of their short-term effects by altering the pH of the wall. Extension is accompanied by turnover of certain wall components; this may be related to the control of growth. It has been proposed that proteins involved in the breakage of hydrogen bonds between cellulose microfibrils and xyloglucan polysaccharides are responsible for wall extension. Longer-term changes in growth rate involve changes in wall synthesis and in the cross-linking of cell wall polymers.

References

Aldington, S., McDougall, G.J. and Fry, S.C. (1991) Structure–activity relationships of biologically-active oligosaccharides. *Plant Cell Env.*, **14**, 625–636.

Carpita, N.G. and Gibeaut, D.M. (1993) Structural models of primary cell walls in flowering plants: consistency of molecular structure with the physical properties of the walls during growth. *Plant J.*, **3**, 1–30.

Cleland, R.E. (1959) Effect of osmotic concentration on auxin action and on irreversible expansion of the *Avena* coleoptile. *Physiol. Plant.*, **12**, 809–825.

Cosgrove, D.J. (1985) Cell wall yield properties of growing tissues. Evaluation by *in vivo* stress-relaxation. *Plant Physiol.*, **78**, 347–356.

Cosgrove, D.J. (1987) Wall relaxation in growing stems: comparison of four species and assessment of measurement techniques. *Planta*, **171**, 266–278.

Cosgrove, D.J. (1993) Wall extensibility: its nature, measurement and relationship to plant cell growth. *New Phytol.*, **124**, 1–23.

Cosgrove, D.J. and Durachko, D.M. (1994) Autolysis and extension of isolated walls from growing cucumber hypocotyls. *J. Exp. Bot.*, **45**, 1711–1720.

Fry, S.C., Smith, R.C., Renwick, K.F. *et al.* (1992) Xyloglucan endotransglycosylase, a new wall-loosening enzyme activity from plants. *Biochem. J.*, **282**, 821–828.

Fry, S.C., Aldington, S., Hetherington, P.R. and Aitken, J. (1993a) Oligosaccharides as signals and substrates in the plant cell wall. *Plant Physiol.*, **103**, 1–5.

Fry, S.C., York, W.S., Albersheim, P. *et al.* (1993b) An unambiguous nomenclature for xyloglucan-derived oligosaccharides. *Plant Physiol.*, **89**, 1–3.

Hart, J.W. (1988) *Light and Plant Growth*, Unwin Hyman, London.

Hayashi, T., Wong, Y. and Maclachlan, G. (1984) Pea xyloglucan and cellulose II. *Plant Physiol.*, **75**, 605–610.

Hoson, T. and Masuda, Y. (1989) Antibodies and lectins specific for xyloglucans inhibiting auxin-induced elongation of azuki bean epicotyls, in *Book of Abstracts and Programme*, (eds S.C. Fry, C.T. Brett and J.S.G. Reid), Proceedings of Fifth Cell Wall Meeting, Scottish Cell Wall Group, Edinburgh, abstr. 132.

Hoson, T., Masuda, Y. and Nevins, D.J. (1992) Comparison of the outer and inner epidermis of auxin-induced elongation of maize coleoptiles by glucan antibodies. *Plant Physiol*, **98**, 1298–1303.

Inouhe, M. and Nevins, D.J. (1991) Auxin-enhanced autohydrolysis in maize coleoptile cell walls. *Plant Physiol.*, **96**, 285–290.

Inouhe, M., Yamomoto, R. and Masuda, Y. (1984) Auxin-induced changes in the molecular weight distribution of cell wall xyloglucans in *Avena* coleoptiles. *Plant and Cell Physiol.*, **25**, 1341–1351.

Kutschera, U. and Schopfer, P. (1986) Effect of auxin and abscisic acid on cell wall extensibility in maize coleoptiles. *Planta*, **167**, 527–535.

Labavitch, J.M. and Ray, P.M. (1974) Turnover of cell wall polysaccharides in elongating pea stem segments. *Plant Physiol.*, **53**, 669–673.

Lockhart, J.A. (1965) An analysis of irreversible plant cell elongation. *J. Theoret. Biol.*, **8**, 264–275.

McDougall, G.J. and Fry, S.C. (1989) Structure–activity relationship for xyloglucan oligosaccharides with anti-auxin activity. *Plant Physiol.*, **89**, 883–887.

McQueen-Mason, S., Durachko, D.M. and Cosgrove, D.J. (1992) Two endogenous proteins that induce cell wall extension in plants. *The Plant Cell*, **4**, 1425–1433.

McQueen-Mason, S., Fry, S.C., Durachko, D.M. and Cosgrove, D.J. (1993) The relationship between xyloglucan endotransglycosylase and in-vitro cell wall extension in cucumber hypocotyls. *Planta*, **190**, 327–331.

Nishitani, K. and Masuda, Y. (1983) Auxin-induced changes in cell wall xyloglucans: effects of auxins on the two different subfractions of xyloglucans in the epicotyl cell walls of *Vigna angularis* [azuki bean]. *Plant and Cell Physiol.*, **24**, 345–355.

Pavlova, Z.N., Ash, O.A., Vnuchkova, V.A. *et al.* (1992) Biological activity of a synthetic pentasaccharide fragment of xyloglucan. *Plant Sci.*, **85**, 131–134.

Rayle, D.L. and Cleland, R.E. (1992) The acid growth theory of auxin-induced cell elongation is alive and well. *Plant Physiol.*, **99**, 1271–1274.

Revilla, G., Sierra, M.V. and Zarra, I. (1986) Cell wall autohydrolysis in *Cicer arietinum* L. epicotyls. *Journal of Plant Physiol.*, **122**, 147–157.

Seara, J., Nicolas, G. and Labrador, E. (1988) Autolysis of the cell wall, its possible role in endogenous and IAA-induced growth in epicotyls of *Cicer arietinum*. *Plant Physiol.*, **72**, 769–774.

Talbot L.D. and Ray, P.M. (1992) Molecular size and separability features of pea cell wall polysaccharides. Implications for models of primary wall structure. *Plant Physiol.*, **92**, 357–368.

Wakabayashi, K., Yamaura, K., Sakurai, N. and Kuraishi, S. (1993) Unchanged molecular weight distribution of xyloglucans in outer tissue cell walls along intact growing hypocotyls of squash seedlings. *Plant and Cell Physiol.*, **34**, 143–149.

Further reading

Cosgrove, D.J. (1993) How do plant cell walls extend? *Plant Physiol.*, **102**, 1–6.

Preston, R.D. (1982) The case for multinet growth in the growing walls of plant cells. *Planta*, **155**, 356–363.

Taiz, L. (1984) Plant cell expansion: regulation of cell wall mechanical properties. *Ann. Rev. Plant Physiol.*, **35**, 585–657.

6 The cell wall and intercellular transport

6.1 Symplastic and apoplastic transport

All transport of materials between plant cells involves the cell wall. Such transport can occur in two main ways (Figure 6.1.). One is **apoplastic transport**, which involves movement through the matrix of the cell wall (the **apoplast**) and may occur between neighbouring cells (in which case the distance moved through the wall will be only a few microns) or between widely separated cells. Apoplastic transport involves not only movement through the cell walls but also movement across the plasma membrane, if transport begins or ends within the protoplast. The second form of transport is **symplastic transport**, which involves movement of material from one cell to another via the plasmodesmata. Here, the cell wall is involved in as far as the plasmodesmata may be considered to be an integral part of the cell wall.

Apoplastic transport is the type which is most intimately involved with the molecular structure of the cell wall. The presence of relatively immobile microfibrils and matrix polymers in the wall means that the wall behaves like a sieve, and large molecules and microorganisms are unable to pass through it. The aqueous channels through the matrix do, however, permit the passage of small molecules and ions, including some small proteins and polysaccharides. It is not clear what the upper limit of size is for movement through the apoplast; it is probably around 10^1 daltons, but some larger proteins and proteoglycans do appear to penetrate the wall (section 3.5.3). Ease of movement does not depend on molecular size alone, since charge will also have an effect (positively charged molecules tend to bind to the predominantly negatively charged wall polymers and are therefore retarded), as will any tendency to bind to wall polymers by non-ionic interactions. For small, uncharged molecules, such as sucrose, the unlignified wall offers very little resistance to movement, and the plant growth substances also move freely, being small and either neutral or negatively charged.

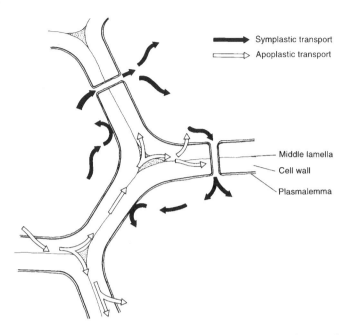

Figure 6.1   Apoplastic and symplastic transport between cells. Note that apoplastic transport allows movement not only through the cell walls, but also across the plasma membrane.

Primary walls thus provide a ready route for the transport of certain molecules. However, the plant regulates this movement by means of certain structures which limit transport and others which enhance it. These are discussed in the remainder of this chapter.

6.2 Adaptations for impermeability

6.2.1 The cuticle

The plant cuticle is a layer of relatively impermeable material on the outside surface of the outer layer of cells (the epidermis) of young, unsuberized tissues (Figures 6.2, 6.3 and 6.4). The cuticle consists principally of **cutin**, a hydrophobic material that is also found as an impregnation of the epidermal cell wall immediately beneath the cuticle. Cutin is a complex mixture of fatty acids and fatty esters. The fatty acids are mostly hydroxylated C_{16} and C_{18} straight-chain saturated molecules, and they are often extensively esterified to each other to form a cross-linked network (Figure 6.5). Within and on the outside of the cuticle there are also discrete layers of wax, containing long-chain hydrocarbons. The cuticle acts as an effective barrier to the movement of water vapour, oxygen and carbon dioxide across the plant/air interface, which means that transpiration and gas exchange occur principally via the stomata. It also means that substances applied to the plant in aqueous solution are rather poorly taken up by the plant, unless the substances concerned can penetrate the cuticle, either by virtue of amphipathic properties or with the aid of detergents.

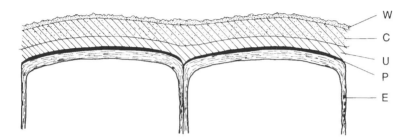

Figure 6.2   A hypothetical cuticle. W, surface wax; C, cuticularized layer consisting of cutin embedded in wax; U, cutinized layer consisting of cutin; P, layer of pectin; E, epidermal cell wall.

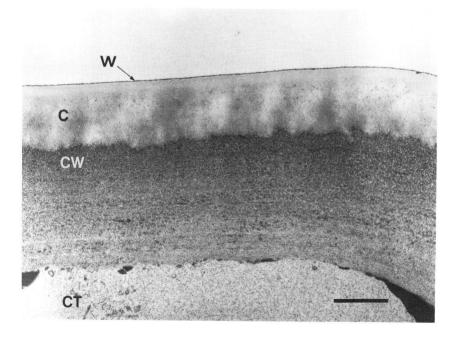

Figure 6.3   TEM through the outer epidermal cell wall of orange leaf exhibiting a heavily-developed cuticle. C, cuticle; CW, cell wall; CT, cytoplasm; W, wax. Bar, 1 μm.

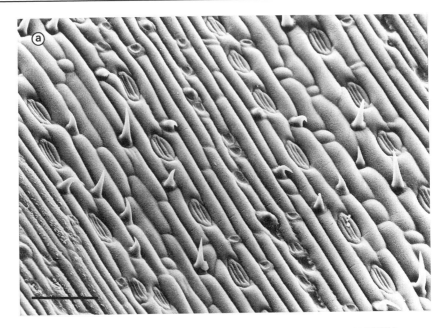

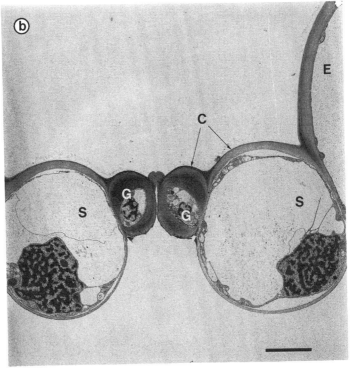

Figure 6.5   Representation of the polymers and their linkages found in cutin.

6.2.2 The Casparian strip

The absorbing regions of roots contain a tissue, the **endodermis**, which surrounds the central vascular cylinder (the **stele**) and separates the stele from the **cortex** (Figures 6.6 and 6.7a). The endodermal cylinder is one cell thick, and the radial and transverse walls contain an impermeable band extending right round the cell, the **Casparian strip** (Figure 6.7a,b,c). This wall region is heavily impregnated with **suberin**, a hydrophobic material related to cutin. This impregnation prevents water and dissolved solutes passing across the endodermis through the radial and transverse walls, i.e. apoplastically. Since a major activity of roots is the uptake of ions and other dissolved materials from the soil, the movement of these substances into the stele (along which they are transported to the remainder of the plant) must occur via the symplast. The plasma membrane of the epidermal cells adheres closely to the wall at the region of the Casparian strip, and there is no path for water movement between the wall and the plasma membrane (Figure 6.6). Hence the amount of a solute entering the stele can be controlled by the necessity for it to cross a plasma membrane from the apoplast into the symplast. The flow of water through the cortex is predominantly apoplastic, so it is at the plasma membrane of the endodermal cells that most of the control is exerted. Once inside the endodermis, the water and solutes can proceed either apoplastically or symplastically to the tracheary elements.

The suberin of the Casparian strip, which is similar to the suberin of cork, consists of long-chain fatty acids and esters similar to those in

Figure 6.4   (left) Cuticle of wheat flagstaff. (a) SEM of the leaf surface showing the cuticle that covers the leaf surface. Note the hairs and stomata. Bar, 100 μm. (b) TEM of epidermis (same tissue) showing the continuous cuticle covering both the stomatal cells as well as the epidermal cells. C, cuticle; S, subsidiary cell; G, guard cell; E, epidermal cell. Bar, 2 μm.

(a)

see (b)

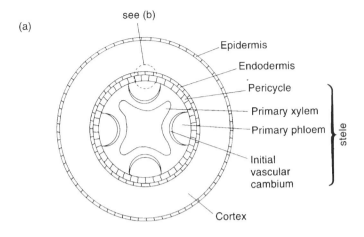

Epidermis

Endodermis

Pericycle

Primary xylem

Primary phloem

Initial vascular cambium

stele

Cortex

(b)

Outside of root

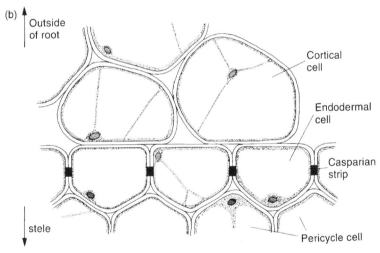

Cortical cell

Endodermal cell

Casparian strip

stele

Pericycle cell

(c)

Outside of root

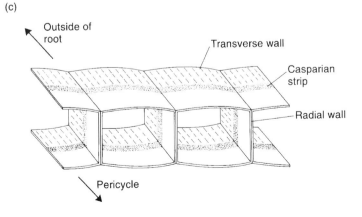

Transverse wall

Casparian strip

Radial wall

Pericycle

cutin (section 6.2.1) but slightly longer on average (up to C_{22}). They are often esterified to phenols, such as ferulic acid (Figure 6.8).

6.2.3 Lignin

Lignification of cell wall results in a great decrease in the wall's permeability to water and to dissolved solutes. The lignin replaces water throughout the bulk of the wall matrix, leaving a few narrow pores through which water can penetrate (Figure 6.9).

6.3 Special adaptations for transport

6.3.1 Plasmodesmata

Plasmodesmata (singular: plasmodesma) provide a cytoplasmic connection linking adjacent cells through their common cell wall (Figure 6.10). They are thin, irregular cylinders of cytoplasm, 30–60 nm in diameter, bounded by plasma membrane which lines the surface of the wall. Through the middle of the plasmodesma runs a tube of membrane, the **desmotubule**, which appears to be continuous with the endoplasmic reticulum of the overlying cells. A central rod may run down the centre of the desmotubule. There is some evidence for a ring of protein surrounding each end of the plasmodesma, between the plasma membrane and the wall. This ring of protein might act as a sphincter, limiting movement through the plasmodesma, but this idea is unproven as yet.

Plasmodesmata follow irregular paths through the wall and are sometimes branched (Figure 6.11). There is often a widening of the plasmodesma as it passes through the middle lamella, and branching may take place at this point. The number of plasmodesmata linking neighbouring cells is quite variable, from fewer than one to more than 15 per μm^2 of wall area, and different walls of the same cell may have different densities of plasmodesmata.

Small molecules and ions pass readily through the plasmodesmata. Each plasmodesma may contain two separate channels for transport between neighbouring cells: a **cytoplasmic annulus** between the plasma

Figure 6.6   (left) The endodermis. (a) Diagram of root tissue in transverse section showing the position of the endodermis. (b) Magnified portion of the root section encircled in (a) showing the strategic position of the Casparian strip in the endodermal cells. (c) Three-dimensional view of two endodermal cells. The dotted area represents the suberized Casparian strip. Water from solution outside the root wets the cellulose walls until it reaches the Casparian strip (broken-line-shaded areas). Since water cannot penetrate the hydrophobic Casparian strip, the cell walls inside the Casparian strip are wetted by water that has passed through the protoplast of the endodermal cell.

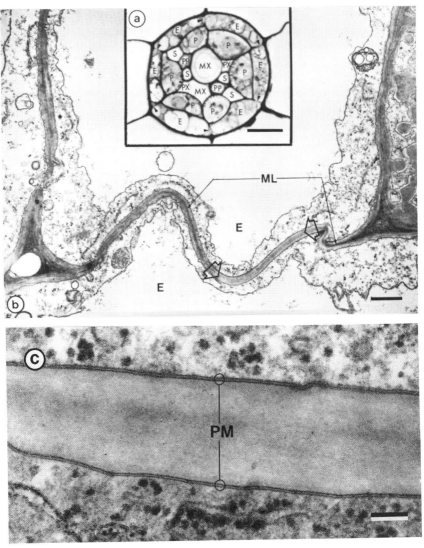

Figure 6.7   Endodermis and Casparian strip. (a) TS of the central stele (vacuolar tissue) of an *Azolla* root. Cell types: endodermal cell (E), pericycle cell (P), sieve element (S), phloem parenchyma (PP), protoxylem (PX), metaxylem (MX). Arrow heads point to the Casparian strips. Bar, 10 µm. (b) The radial wall separating two endodermal cells (E) runs across the bottom of this micrograph. Part of a cortex cell is seen at the left, and part of a pericycle transfer cell at the right. The Casparian strip lies between the open arrows. Even at this low magnification it can be seen to have a·more homogeneous texture than the neighbouring wall. For instance the middle lamella (ML) almost disappears in the strip. Notice too that the undulations of the plasma membrane become smoothed out at the Casparian strip. Tissue as for (a). Bar, 1 µm. (c) A portion of Casparian strip at higher magnification. The smooth texture of the strip is probably due to the impregnation of the wall with cutin or suberin. Again, note the smoothness of the plasma membrane (PM) against the strip. It contrasts with the plasma membrane of a normal cell wall (d).

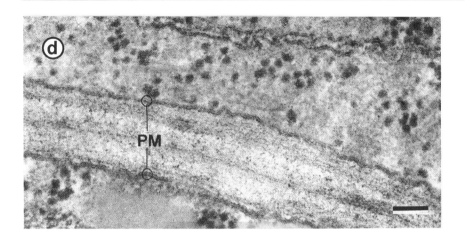

Figure 6.8   Representation of the polymers and their linkages found in suberin.

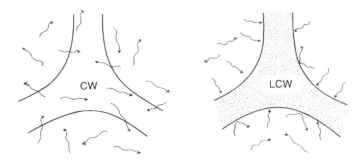

Figure 6.9   Impedance of water movement by lignification. LCW, lignified cell wall; CW non-lignified cell wall.

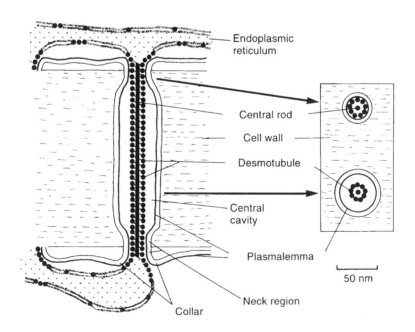

Figure 6.10   A simple plasmodesma. This diagram shows the various plasmodesmatal features seen in electron micrographs. It is not meant to imply any general uniformity of structure, nor specific features.

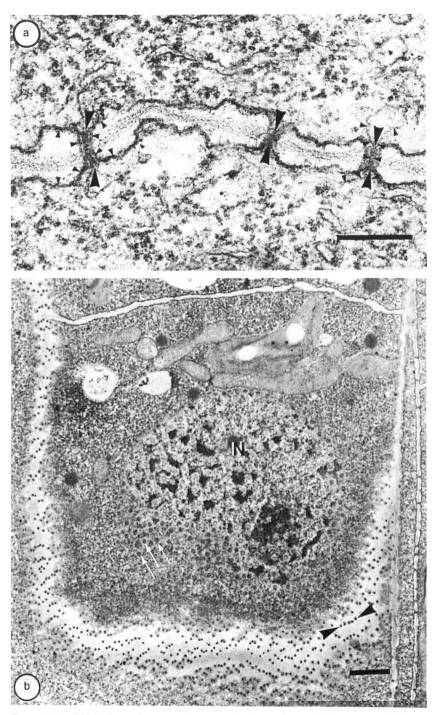

Figure 6.11 TEMs illustrating several aspects of plasmodesmatal ultrastructure. (a) Plasmodesmata piercing the wall between two meristematic cells in a cabbage root. Small arrows show the plasma membrane. Large arrows show the desmotubules. Bar,

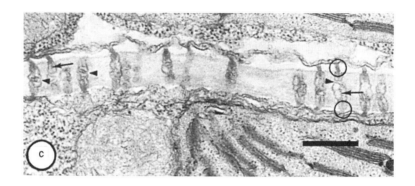

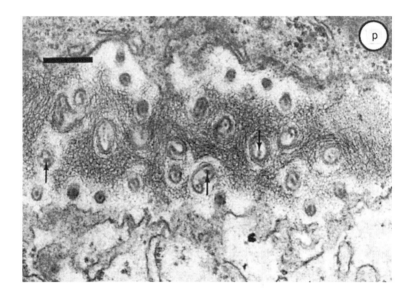

500 nm. (b) End-on views of very numerous plasmodesmata in a section that grazes a curved cell wall in the stele of an *Azolla* root meristem. N, nucleus. Bar, 1 μm. (c) and (d) Plasmodesmata grouped in primary pit fields. Arrows indicate axial desmotubules. (c) Oat leaf. Circles show regions in which the plasmamembrane is closely constricted around the desmotubules at the cytoplasmic extremities of the plasmodesmata. Arrow heads indicate strange convolutions in the central parts of the plasmodesmata. Bar, 500 nm. (d) Pea leaf. Bar, 200 nm. (e)–(g) Compound plasmodesmata in which several canals meet in the interior of the wall. (e) Lupin leaf transfer cell. Bar, 500 nm. (f) Vascular parenchyma of *Polemonium* stem. Bar, 200 nm. (g) Lupin stem. Bar, 500 nm.

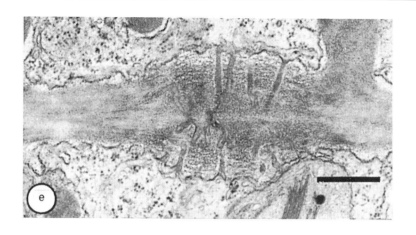

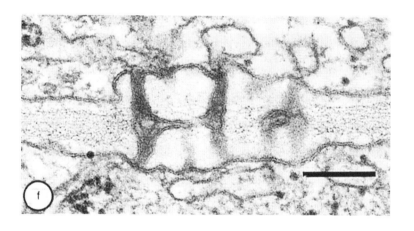

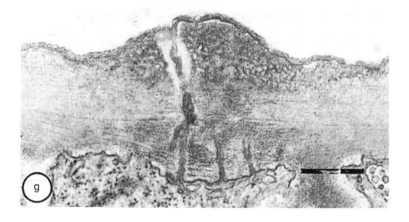

membrane and desmotubule, and a channel through the centre of the desmotubule that links the lumens of the two systems of endoplasmic reticulum. However, the latter channel would be very narrow, and is unlikely to be a major transport route. The upper size limit for transport through plasmodesmata is generally around 700–900 Da. Increases in cytosolic calcium concentration greatly reduce the permeability of the plasmodesmata, and the effect is reversed when calcium levels decline. Thus transport through plasmodesmata may be regulated by calcium. The effects of calcium might be due to increased deposition of callose in the neck region, since calcium stimulates callose synthesis (section 4.2.3). However, no increase in aniline blue staining (for callose) is seen, so calcium is more likely to be acting on the protein in the neck region.

Plasmodesmal permeability can increase under certain circumstances. Plasmolysis breaks plasmodesmata but most reform when the tissue is returned to an isotonic or hypotonic medium. However, the reformed plasmodesmata have a considerably higher exclusion limit (up to 1700 Da) and calcium no longer causes a reduction in permeability. Some viruses also appear to increase plasmodesmatal permeability, permitting movement of virus particles between cells in infected plants.

6.3.2 Pits

Pits occur in the secondary wall where plasmodesmata are present in the underlying primary wall. No secondary wall is laid down immediately over the plasmodesmata, so cytoplasmic continuity between adjacent cells is preserved. The cavities in the secondary wall arise on each side of a group of plasmodesmata, giving rise to a **pit-pair**. If the cavity is of roughly constant diameter all the way through the secondary wall, it is known as a **simple pit** (Figure 6.12a). Often, however, the secondary wall overarches the plasmodesmata to form a **bordered pit** (Figure 6.12b,c). The region of primary wall through which the plasmodesmata pass is called the **pit membrane** in gymnosperms; this membrane may be thickened at the centre to form a **torus**, surrounded by a thinner **margo**. The margo is flexible, and when a large difference in pressure occurs across the pit-pair, the torus is forced against the border, effectively sealing the pit. Such sealed pits are said to be 'aspirated', and can prevent damage to one cell affecting neighbouring cells. This is especially important in the xylem, where damage to one tracheid, followed by loss of tension and sometimes dehydration, is prevented from causing similar damage to neighbouring tracheids.

Under normal circumstances, pits are the means by which water moves from one tracheid to another in the xylem. This occurs principally between adjacent tracheids in a longitudinal column, through the end-wall, but lateral movement may also occur between tracheids, through pits in the side-walls. Such pits also occur in the side-walls of vessel elements, permitting sideways movement of water between vessels (Figure 6.12b).

Both tracheids and vessels lack a protoplast at maturity, so that the plasmodesmata lose their membranous structures (the plasma membrane and desmotubule) and become simply pores through the primary wall. They may be widened by dissolution of some of the matrix materials in the surrounding wall, and the margo of a pit membrane may be reduced to a network of cellulose fibrils, almost denuded of matrix, and hence considerably more permeable than the original group of plasmodesmata.

6.3.3 Phloem sieve plates

Phloem sieve cells communicate via specialized wall regions known as **sieve plates** (Figure 6.13). These regions occur in the end-walls, linking longitudinally adjacent sieve cells, and to a lesser extent also in the side-walls. The pores in a sieve plate may vary in size from a few tens of nanometres (the size of plasmodesmata) up to several micrometres. The broader pores tend to occur in the end walls, consistent with a predominant flow of material through these walls.

The pores are derived from plasmodesmata, which generally widen during the differentiation of the sieve cell. Each pore is lined with callose, though the amount of callose may be small in fully functional sieve tubes. The amount of callose increases as cells age and become less active in transport, and when dormancy occurs the pores may be completely blocked by callose. Similar increases in callose content occur when sieve cells are damaged, and blocked pores may be seen in electron micrographs if care is not taken to minimize disturbance to living cells in the initial stages of specimen preparation. Callose can be produced remarkably quickly and may act as a means of preventing the spread of damage in the phloem, as the aspiration of bordered pits does in the xylem.

6.3.4 Transfer cells

Transfer cells are parenchyma cells which are specialized for short-distance transfer of solutes between cells. Part of the wall of a transfer cell is characterized by irregular ingrowths of wall into the space normally occupied by the protoplast. These ingrowths greatly increase the surface area of the protoplast, and since they are lined by plasma membrane they greatly increase the area of the plasma membrane (Figure 6.14). This is thought to increase the area of membrane available for

(a)

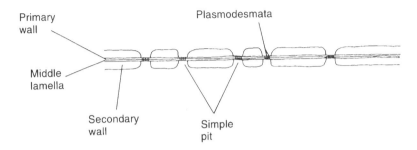

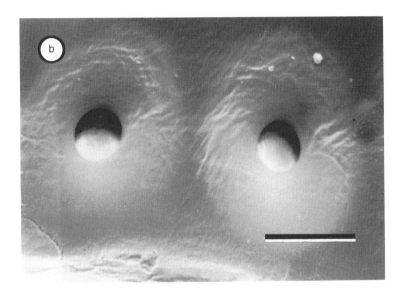

Figure 6.12   Pits. (a) Simple pit showing primary pit fields. (b) SEM of bordered pits in xylem vessels of pine. Bar, 10 μm. (c) As for (b), but with the nearer secondary wall removed. Bar, 10 μm. (d) Diagram of a bordered pit showing the modified secondary wall.

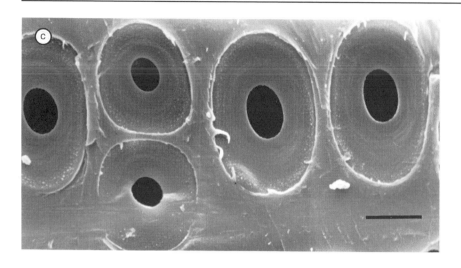

(d)

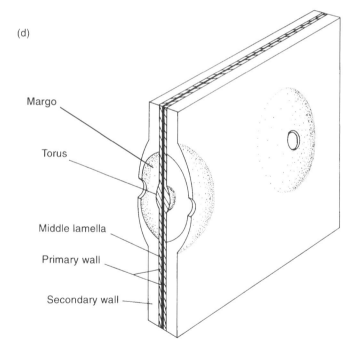

Margo

Torus

Middle lamella

Primary wall

Secondary wall

Figure 6.12

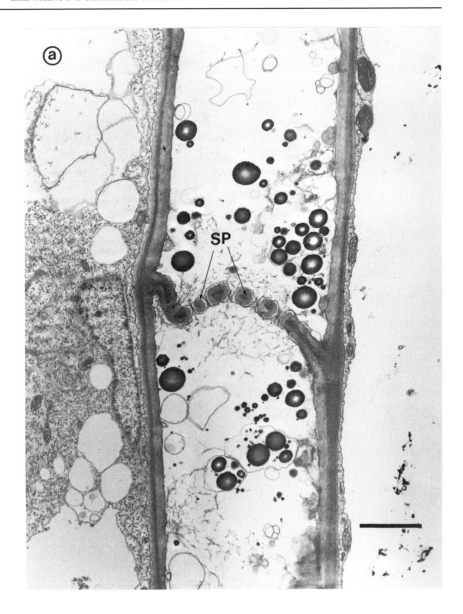

Figure 6.13   TEMs of phloem sieve plates. (a) In LS, sieve tubes (from *Echinochloa colonum* (L.)) are interrupted at intervals by perforated cross walls, the sieve plates (SP), which were originally the walls between successive cells in the file of differentiating sieve elements. The pores develop from plasmodesmata in the primary cross walls. Bar, 2 μm. (b)–(d) Sections through sieve plates from white lupin illustrating the changes that occur during wounding. In (b), (bar, 1 μm) the sieve plate is seen in what is thought to be its normal state, with very small amounts of callose (C) and P-protein (P) fibrils around the pores. Since sieve tubes are under considerable hydrostatic pressure, due to the osmotic effect of their sugary content, damage to the system results in a sudden release of

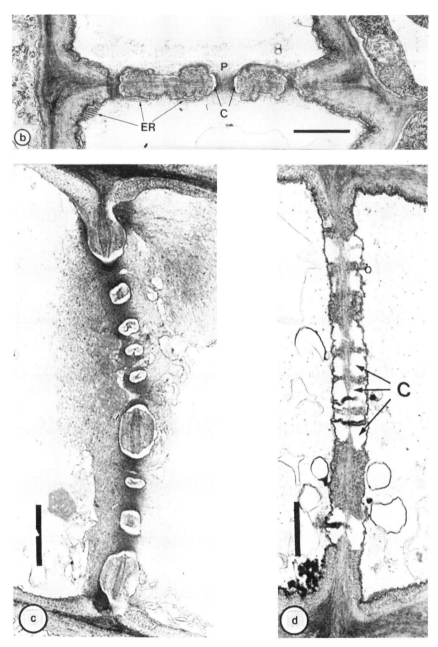

pressure and a surge of liquid down the sieve tube. This carries masses of the P-protein fibrils (normally dispersed throughout the sieve tube) on to the sieve plates and into the pores (c) (bar, 1 μm). A temporary blockage of the pores is thus effected while another reaction occurs – the formation of massive callose plugs which seal the pore (d) (bar, 1 μm). (e) Face view of a sieve plate showing damage response. Tissue from *Coleus blumei* stem. Bar, 1 μm.

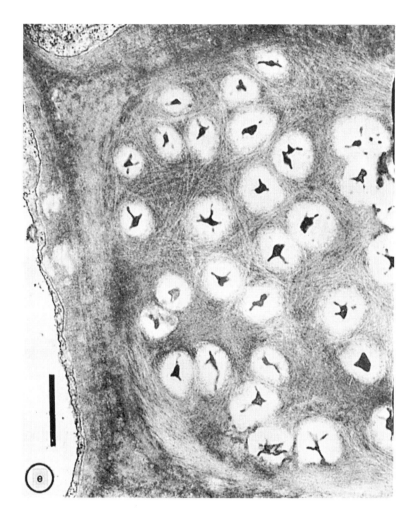

Figure 6.13   (*Continued*)

transport of material into or out of the cell and hence accelerate such transport. The transport is often active and therefore associated with specific pumping proteins in the plasma membrane.

Transfer cells are found characteristically in secretory organs, such as glands and nectaries. They are also found in the xylem and phloem, adjacent to tracheids and sieve cells. Hence they are involved in loading and unloading the vascular systems at source and sink regions of the plant. In the case of phloem, the transfer cells are companion cells, though not all companion cells contain the recognizable wall ingrowths characteristic of transfer cells.

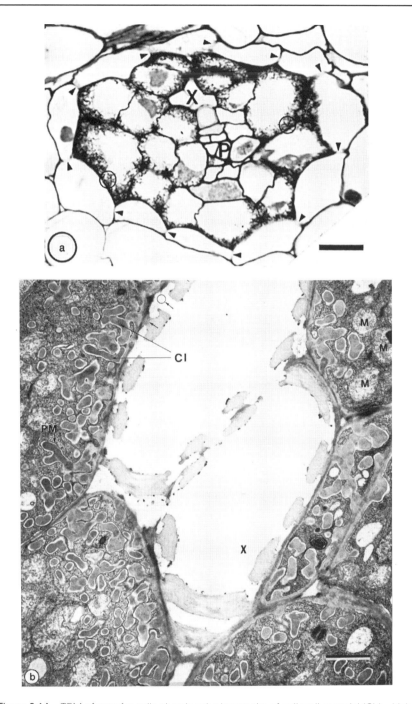

Figure 6.14   TEM of transfer cells showing the ingrowths of cell wall material (CL) which increase the cell-wall surface area. M, mitochondria; X, xylem element; P, phloem. (a) Bar, 20 μm. (b) Bar, 2 μm. Circles in (a) show areas of typical wall ingrowth.

Summary

Apoplastic transport involves movement of molecules through the aqueous part of the cell wall matrix. The cuticle, the Casparian strip and lignified walls are resistant to such movement. Plasmodesmata provide a route for the movement of molecules from cell to cell without crossing the plasma membrane. Pits provide for transport through the secondary wall. Movement between cells in phloem sieve tubes occurs through sieve plates. Transfer cells are specialized for transport of material into and out of the vascular system and secretory organs.

Further reading

Carpita, N., Sabularse, D., Montezumos, D. and Delmer, D.P. (1979) Determination of the pore size of cell walls of living plant cells. *Science*, **205**, 1144–1147.

Kolattakudy, P.E., Espelie, K.E. and Soliday, C.L. (1981) Hydrophobic layers attached to cell walls. Cutin, suberin and associated waxes, in *Encyclopedia of Plant Physiology, New Series*, Springer, Berlin, Vol. 13B, pp. 225–254.

Robards, A.W. and Lucas, W.J. (1990) Plasmodesmata. *Ann. Rev. Plant Physiol. Mol. Biol.*, **41**, 369–419.

7 The cell wall and interactions with other organisms

7.1 Pathogens and potential pathogens

Higher plants are exposed to attack by a wide range of microorganisms which are potentially pathogenic. These include viruses, bacteria and fungi. However, the vast majority of microorganisms are unable to attack higher plants successfully, due to a wide range of defence mechanisms possessed by the plant. These include both passive and active defences, and almost every form of defence involves the cell wall in some way. The subtlety of these interactions provides an insight as to why the cell wall is structurally complex, and has given rise to the current interest in cell wall polysaccharides as possible sources of signals for cell–cell communication.

7.1.1 The wall as a passive barrier

The intact plant cell wall is an extremely effective physical barrier against attack by microorganisms. Its pores are far too small to permit even viruses to penetrate through it to the protoplast (though once within the protoplast, viruses may spread to other cells through plasmodesmata). Penetration therefore requires either enzymic dissolution of part of the wall (Figures 7.1 and 7.2), or opportunistic attack through breaks in the wall. Dissolution of the wall by enzymes from pathogens is considered in more detail in Chapter 9.

Some fungi can parasitize plants without penetrating into the protoplast. Instead, they degrade the middle lamella of the wall, and grow through the host tissue between the cells. Initial penetration is often through stomata (Figure 7.3). They then parasitize the plant by absorbing those low molecular weight materials which normally travel through the apoplast, including sucrose, ions and amino acids. Degradation of the middle lamella requires secretion only of pectinases, in contrast to the wide range of enzymes needed to penetrate the wall through to the protoplast.

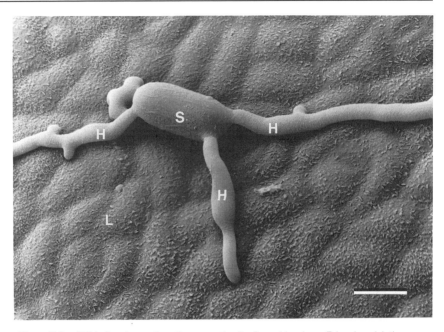

Figure 7.1 SEM showing cell-wall penetration by fungal hyphae. *Erisyphe pisi*, the grey mould of peas, is growing on the surface of a pea leaf, *Pisum sativum*. The spore (S) has produced three hyphal filaments (H). These travel across the surface of the leaf (L) and gain access to the interior either by penetrating a stomatal pore or, as shown here, by dissolution of the cuticle and epidermal cell wall. The rough surface of the leaf is due to the presence of cuticular wax on the specimen. Bar, 30 µm.

Once lignification of the cell wall has occurred, the wall becomes impenetrable to all but a very few fungi (Chapter 9). This applies both to those fungi whose strategy is to pass though the wall into the proto-plast, and to those which penetrate between cells by dissolving the mid-dle lamella, since lignification begins in the middle lamella.

The presence of cutin and, in grasses, silica in the epidermal cell wall and in the cuticle provides an initial barrier to fungal attack. Penetration through the cutin requires secretion of a cutinase, and silicon appears to act either as a mechanical barrier or as a toxin against some fungi.

7.1.2 The wall as an active line of defence

The most important active mode of defence against pathogens that involves the cell wall is the deposition of lignin (and sometimes suberin) as a response to infection. This occurs as part of the **hypersensitive response**, in which cells around the point of parasitic infection rapidly undergo lignification and cell death (Figure 7.4). The point of infection is effectively sealed off from the rest of the plant by a layer of lignified cells, so that the parasite can penetrate no further. The response is triggered by both fungi and bacteria, and successful parasitism is chiefly confined to those microorganisms which avoid triggering the response, and which therefore enter into a pseudosymbiotic relationship with the host plant in the early stages of infection. The triggers which signal the response are considered in the next section.

A more limited active defence mechanism involving the wall does not involve cell death. This is the deposition of callose at the point of penetration of a fungal haustorium through a cell wall. Once such penetration begins to disturb the plasma membrane, callose is rapidly laid down at the inside surface of the wall, adjacent to the plasma membrane (Figure 7.5). The callose provides an additional barrier to further penetration into the protoplast. The deposition of callose is probably triggered by an influx of calcium into the cell, since $\beta(1–3)$glucan synthase (callose synthase) is calcium-dependent. This influx may result directly from disturbance to the plasmamembrane, since the intact membrane actively pumps calcium out of the cytoplasm into the wall, maintaining a large calcium gradient across the membrane. The callose layer, known as a **papilla**, also contains other wall materials, including lignin-like phenolic polymers and cellulose.

A number of proteins accumulate in cell walls in response to pathogenic attack. These include extensin and other HRGPs, together with chitinase and β-glucanase, which may attack fungal cell walls, and proteins which are **polygalacturonase inhibitors** (PGIPs). These PGIPs defend the pectin of the plant cell wall against attack by pathogen-derived polygalacturonase, and also enhance the role of oligogalacturonides as signalling molecules by limiting their further breakdown to shorter, inactive oligosaccharides (see below). All these proteins, together with intracellular proteins produced during microbial attack, may be classed together as **pathogenesis-related proteins** (PRPs). Their synthesis and deposition may be induced by the action of a single signal transduction pathway, or a group of interacting pathways.

7.1.3 The wall as a signalling device

The wall not only provides a physical barrier to attack by pathogens but also participates in the mechanisms by which the active defences are triggered. The best understood example of this is in the elicitation of phytoalexin production in response to pathogenic attack.

Phytoalexins are non-specific toxins produced by higher plants when attacked by pathogens. They have widely varying structures (Figure 7.6), but they probably act by disrupting membranes, and relatively high concentrations of phytoalexins are produced in order to achieve this. They are toxic not only to microorganisms but also to higher plant tissues;

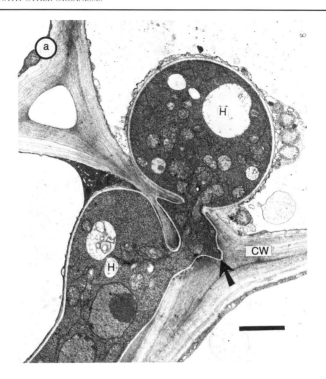

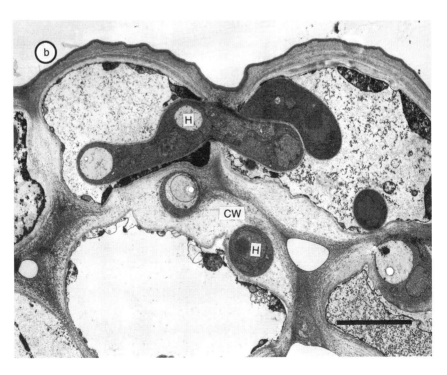

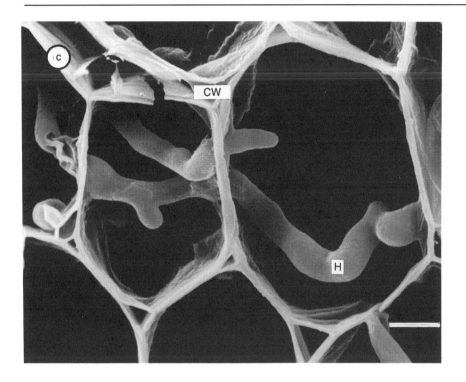

Figure 7.2   Electron micrographs showing penetration of cell walls of *Phaseolus vulgaris* by the pathogenic fungus *Colletotrichum lindemuthianum*. (a) Early, benign phase of infection: TEM showing penetration of cell wall (CW), by hypha (H), which has filled an intercellular space (arrow). Note that cell-wall dissolution is highly localized and that the upper cell is dead. Bar, 2 μm. (b) Later, destructive phase of infection: TEM showing hyphae (H) penetrating and developing within host cell walls (CW). Note extensive dissolution and swelling of the wall. Bar, 5 μm. (c) Destructive phase of infection: SEM showing hyphae (H) penetrating cell walls (CW) and crossing lumens of cortical cells. Bar, 5 μm.

however, this does not disadvantage the host, since phytoalexin production is part of a defensive system which often includes necrosis of host tissues.

The production of phytoalexins is triggered by the action of factors known as **elicitors**. These include both exogenous elicitors, produced by pathogens, and endogenous elicitors, produced by higher plants in response to a variety of stresses. These stresses include not only pathogenic attack but also abiotic stresses such as extremes of temperature and mechanical damage. A variety of different types of compound can act as elicitors, including oligosaccharides produced both from fungal wall polysaccharides and from host wall polysaccharides.

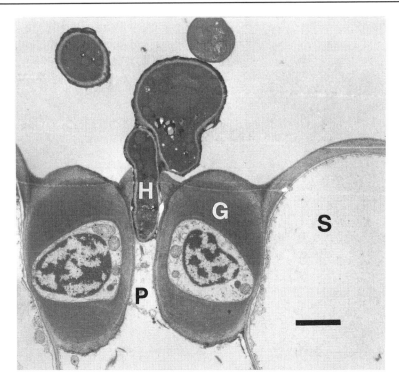

Figure 7.3 TEM of a fungal hypha (H) penetrating a stomatal pore (P) on the surface of a wheat leaf. G, guard cell; S, subsidiary cell. Bar, 2 μm.

The best-known elicitors derived from fungal walls are non-cellulosic glucans. They have been studied in detail in the interaction between soybean (*Glycine max*) and the fungus, *Phytophthora megasperma*. The soybean cell wall contains a β-glucanase which attacks the cell walls of the fungus as the pathogen invades the host tissue. The fungal wall polysaccharide which it degrades is a β(1–6)glucan with β(1–3) side-chains, and some of the resulting oligosaccharides and polysaccharides are potent elicitors. Similar degradation products are produced by the action of mild acid on the cell walls, and the elicitor activity of these acid-hydrolysis products has been studied in detail. One particular fragment, a heptasaccharide, is a potent elicitor. It has a highly specific structure (Figure 7.7), and any variation in the structure abolishes its activity as an elicitor. For instance, changing the anomeric configuration at any point abolishes activity, as does changing the position of either of the β(1–3)-linked glucose residues. The heptasaccharide can, however, be extended at its reducing end without losing all activity.

Figure 7.4   Hypersensitive response. (a) The effects of *Xanthomonas campestris* pv. *campestris* and *Xanthomonas campestris* pv. *vitians* on *Brassica campestris* leaves 24 h after inoculation. The left hand side of the leaf was inoculated with *X. c. campestris* (strain 8004) in two places and the right hand side with *X. c. vitians* (strain 9000) in three places with 10⁷ cells per ml. (b) SEM of TS through healthy leaf tissue of *Brassica campestris* showing

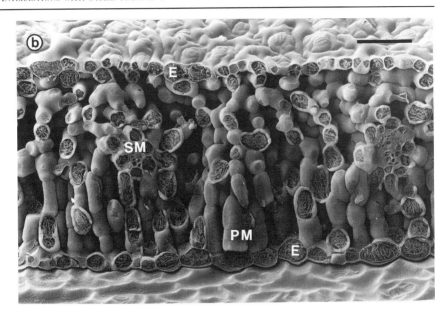

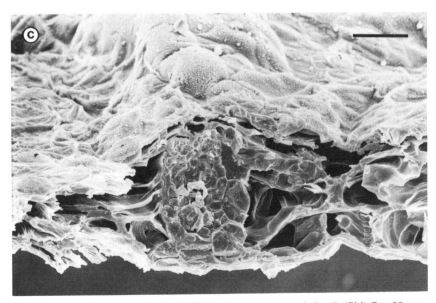

epidermis (E), palisade mesophyll cells (PM) and spongy mesophyll cells (SM). Bar, 30 μm. (c) SEM of TS through leaf tissue of *Brassica campestris* exhibiting a hypersensitive response due to infection by *Xanthomonas campestris* pv. *vitians*. This response is characterized by tissue collapse, loss of membrane integrity, vein blockage and melanin production. Bar, 30 μm.

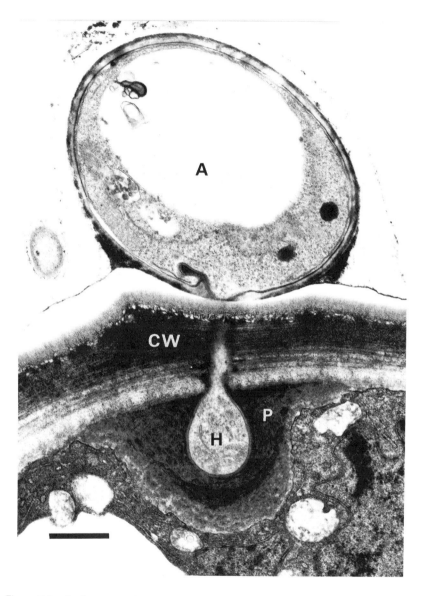

Figure 7.5   Resistant reaction: TEM showing encasement of intracellular hypha (H) of the pathogenic fungus *Colletotrichum lindemuthianum* by multi-layered wall apposition or papilla (P) produced by an epidermal cell of *Phaseolus vulgaris*. Such hyphae do not develop further. The penetration of the epidermal cell wall (CW) is not associated with extensive wall dissolution. A, appressorium. Bar, 1 μm.

$$CH_3CH{=}CH(C{\equiv}C)_3CH{=}CHCH(OH)CH_2OH$$

safynol

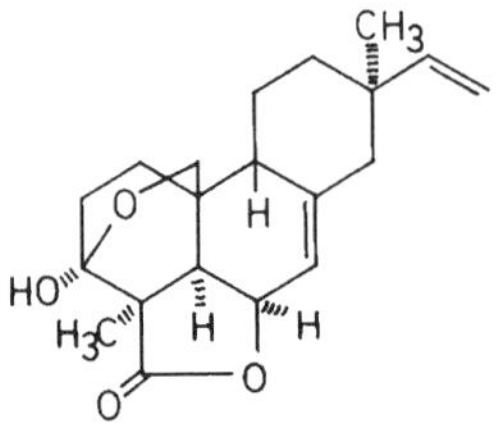

momilactone B

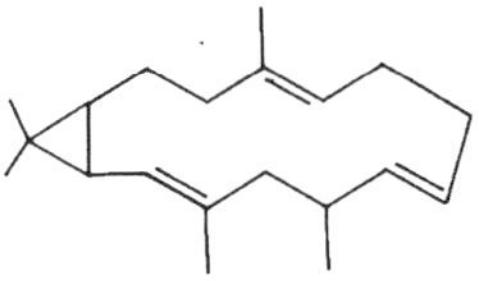

casbene

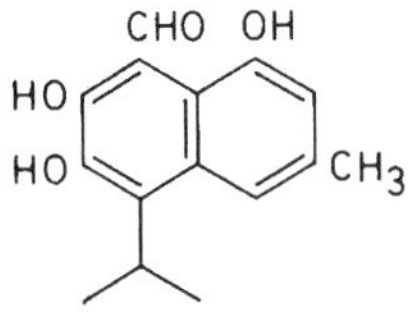

hemigossypol

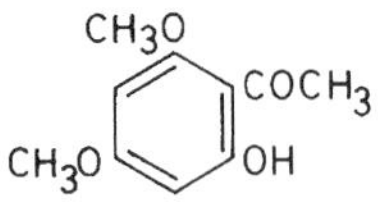

xanthoxylin

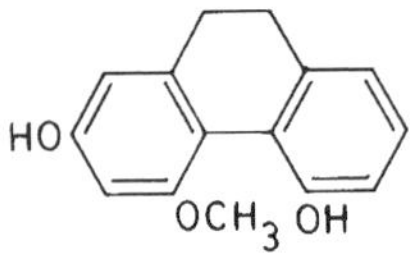

hircinol

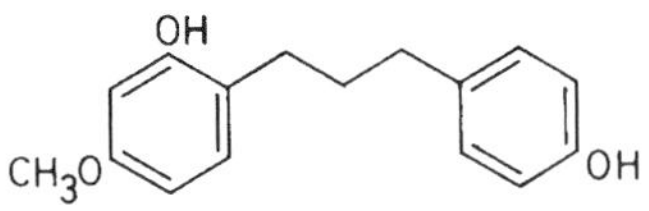

broussonin A

Figure 7.6 A range of phytoalexin molecules produced by different plants in response to infection by pathogenic organisms.

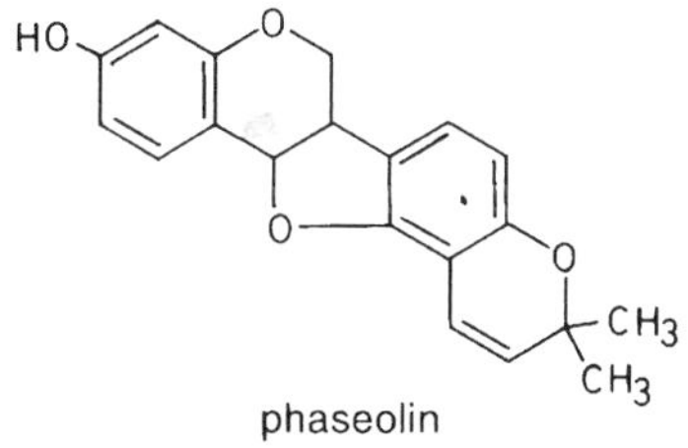

phaseolin

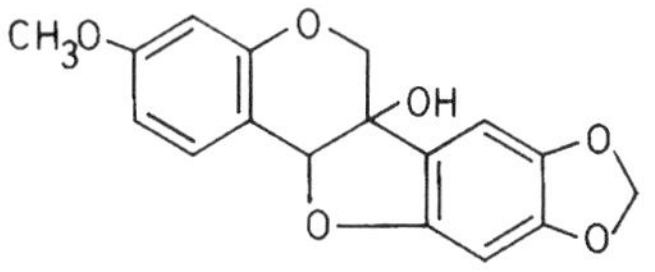

pisatin

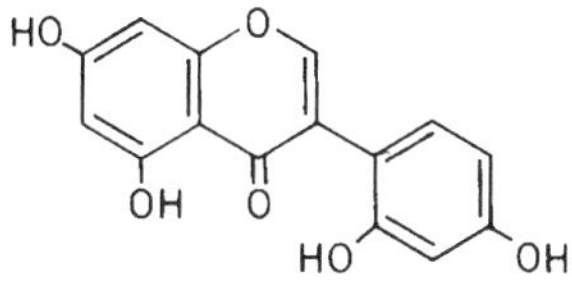

2' hydroxygenistein

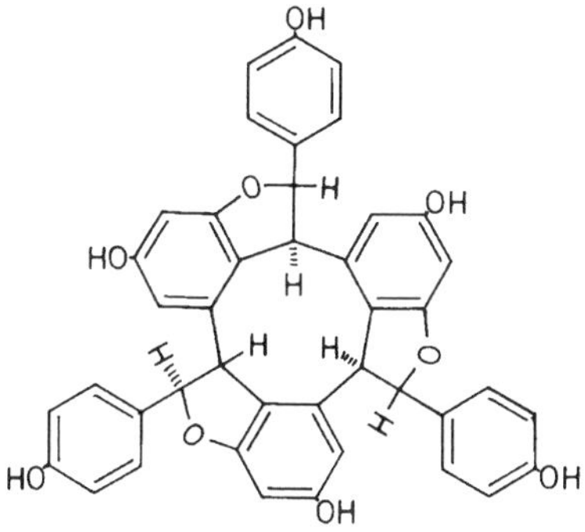

α viniferin

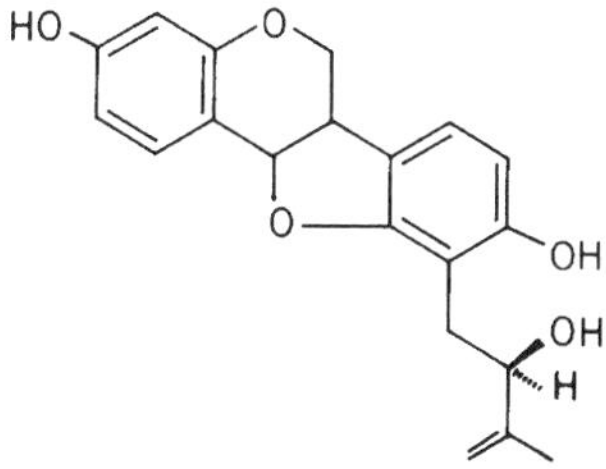

(−) – dolichin A

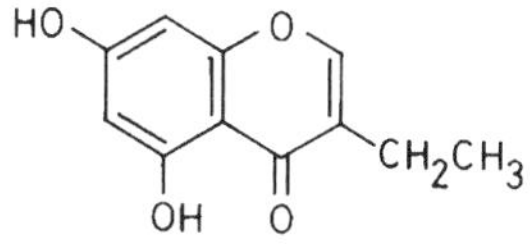

lathodoratin

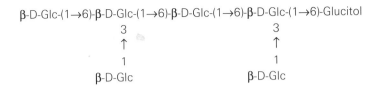

β-D-Glc-(1→6)-β-D-Glc-(1→6)-β-D-Glc-(1→6)-β-D-Glc-(1→6)-Glucitol

Figure 7.7 A heptasaccharide phytoalexin elicitor produced as a result of infection of *Glycine max* by the fungus *Phytophthora megasperma*. The elicitor is a fragment of the fungal wall produced by the activity of *G. max* enzymes.

More recently it has become clear that just as a host enzyme acts on the fungal wall to produce an exogenous elicitor, so a fungal enzyme acts on the host cell wall to produce an endogenous elicitor (Figure 7.8). *Phytophthora* secretes an endopolygalacturonase as part of its offensive armoury. This enzyme acts on homogalacturonan in the host wall to produce oligogalacturonans which are active elicitors. Such oligogalacturonans can be produced from cell wall preparations by partial acid hydrolysis, and in these partial hydrolysates activity resides in $\alpha(1–4)$-linked oligomers 10–13 residues long, with the dodecagalacturonan (12 residues) being the most active. The oligogalacturonans not only act as elicitors themselves but also exhibit synergism when present together with oligoglucan exogenous elicitors. Oligogalacturonan elicitors are now known to be produced by many plants in response to a wide range of pathogens, including both fungi and bacteria.

Phytoalexin production is not the only defence reaction that is triggered by host cell wall fragments. The hypersensitive response is also induced by wall polysaccharide fragments produced by partial acid hydrolysis of wall preparations. Like the endogenous elicitors of phytoalexin production, these necrotic factors probably originate from the pectic polymers. However, their production in host–pathogen interactions has not yet been demonstrated.

Another defence mechanism that is activated in response to infection is the production of inhibitors of microbial proteinases. These proteinases degrade host proteins, and the inhibitors are thought to protect the host from this. Mechanical injury causes plants to synthesise these proteinase inhibitors, and tissues remote from the site of injury are reported to form the inhibitors in response to a signal emanating from the injured tissues. This signal has been termed **proteinase inhibitor inducing factor** (PIIF). There is evidence that fragments of pectic polysaccharides have PIIF activity and, like the oligosaccharins inducing phytoalexin activity, the active fragments concerned appear to be short chains of $\alpha(1–4)$-linked galacturonic acid residues. These oligogalacturonides are not, however, mobile in plant tissues and therefore cannot be responsible for the long-distance signals which stimulate pro-

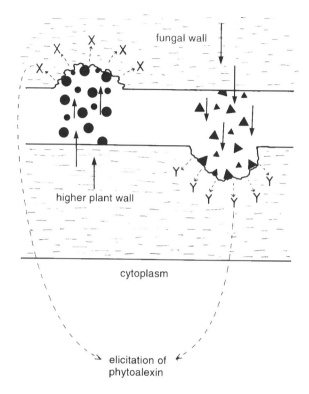

Figure 7.8 Production of elicitor fragments from the cell walls of both the pathogen and host by the activity of enzymes of one acting on the cell wall of the other. Circles, higher plant enzyme; triangles, fungal enzyme. X, elicitor derived from fungal wall; Y, elicitor derived from higher plant cell wall.

teinase inhibitor synthesis at remote sites. Oligogalacturonides are also able to induce the formation of the PRPs described in section 7.1.2, together with peroxidase and other enzymes involved in lignin biosynthesis.

Oligogalacturonides thus act as triggers for a number of defence reactions. They do so as calcium-linked dimers, with the 'egg-box' conformation (section 3.2.2). One very early effect of these dimers is plasma membrane depolarization, accompanied by an influx of calcium and protons and an efflux of potassium (Messiaen *et al.*, 1993). The consequent rise in cytosolic calcium concentration may be part of the signal transduction pathway leading to the observed defence reactions. Phosphorylation of plasma membrane proteins has also been observed in response to oligogalacturonide treatment (Jacinta *et al.*, 1993). No plasma membrane-binding protein for oligogalacturonides has been reported, though a plasma membrane protein which binds the *Phytophthora megasperma* heptasaccharide elicitor has been found in soybean. Fungal elicitors also appear to inhibit the H^+-ATPase in some cells, and to bring about membrane depolarization.

While oligogalacturonides are now well established as endogenous signalling molecules in the defence responses of higher plants to pathogens, other cell wall polysaccharides may also act as sources of such signals. There have been suggestions that both xylan oligosaccharides and xyloglucan oligosaccharides may induce the formation of PRPs. Conversely, oligogalacturonides may have cell-signalling effects other than their roles in plant–pathogen interactions. They inhibit auxin-induced growth in the pea stem assay used to study the growth regulating effects of xyloglucan oligosaccharides (Box 5.2), though at much higher (millimolar) concentrations. They also influence morphogenesis in thin cell layers of tobacco (Aldington *et al.*, 1991). A related wall fragment, rhamnogalacturonan II (section 2.5.6), may also have cell-signalling activity (Aldington and Fry, 1994).

7.1.4 The wall as a point of attachment for the pathogen

There is one situation in which the cell wall appears to act as a help, rather than a hindrance, to a pathogen. This is in the infection of dicotyledonous plants by *Agrobacterium tumefaciens*, which causes crown gall disease. The bacterium binds to pectin, and it is likely that this interaction enables the pathogen to adhere to the outer surface of the plant at the first stage of the infection process. A calcium-binding protein, ricadhesin, on the bacterial surface is involved in the adhesion process, as is a vitronectin-like glycoprotein on the plant cell surface (Stacey *et al.*, 1992). Since *Agrobacterium* is an effective vector for the insertion of foreign genetic material into higher plants, these interactions are of considerable importance in plant biotechnology.

7.2 Predators

The cell wall provides opportunities for the plant to defend itself not only against microorganisms but also against larger predators. The deposition of lignin renders plant tissues tough and unpalatable, and higher animals are unable to degrade lignin. The presence of silica in the cell walls and epidermal cell lumens of grasses may also play a part in decreasing the palatability of those plants that contain it.

These passive defences are supplemented by active ones, at least as far as insect attack is concerned. Insects, like microorganisms, produce proteinases which degrade the host plant tissues. The same inhibitors that inhibit microbial proteinases also inhibit the proteinases from insects. Hence the response triggered by PIIF as a result of the mechanical injury caused by insect attack will be effective in inhibiting the action of the insect proteinases on plant proteins.

7.3 Nodulation and nitrogen fixation

There are two major groups of nitrogen-fixing microorganisms which form symbiotic relationships with higher plants. These are the **rhizobia**, which colonize legumes and clover (Figure 7.9), and **actinomycetes** such as *Frankia*, which colonize a variety of plants, including alder. As far as the host cell wall is concerned, the two types of symbiosis are broadly similar. The first contact in each case is between the cell wall of the free-living microorganism and the root hairs of the host. In rhizobia, there is considerable specificity in the interaction; each cultivar of host only interacts with certain strains of rhizobia, and vice versa. Specificity is mediated by a series of signals which pass between the two partners. First, the bacterium detects substances diffusing from the plant root. It is chemotactic towards a variety of nutrients, such as sugars and amino acids, but in addition it is attracted by specific flavonoid compounds produced by the legume. These flavonoids induce the transcription of the *nod* genes of *Rhizobium*. The products of the *nod* genes bring about the formation of a specific lipooligosaccharide in the bacterium; this compound diffuses to the root surface, where it induces root-hair curling. In the case of *Rhizobium melilotii*, it has been fully characterized: it is a sulphated tetrasaccharide of N-acetylglucosamine, with one acetyl group replaced by a di-unsaturated C_{16} fatty acid, and it has been named Nod-Rm1. A further element of specificity may be mediated by interactions between lectins on the root-hair wall and sugars on the surface of the rhizobium. However, this is still a matter of debate.

Once contact has been established, the root hair either curls, in the case of rhizobia (Figure 7.9a–c), or becomes deformed, in the case of actinomycetes. The microorganisms appear to be able to dissolve away a small area of the outer wall of the root hair, and to penetrate into the interior. However, as the microorganism passes into the root hair, the host tissue forms new wall material around it. The microorganism divides to form an infection thread penetrating the tissues of the root, and the host continues to form cell wall around the growing thread (Figure 7.9d–f). This specialized wall material appears to be pectic in nature and remains in intimate contact with the microorganism as the nitrogen-fixing nodule forms and becomes active.

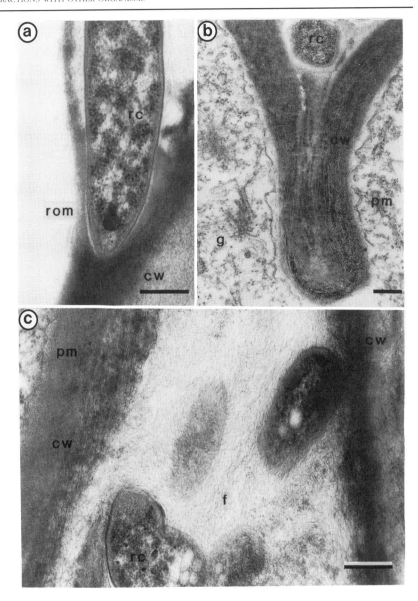

Figure 7.9 Legume–rhizobium symbiosis between root hairs of *Vicia hirsuta* and *Rhizobia leguminosarum* strain 8401pRL1J1. (a) Portion of rhizobial cell embedded in the outer layers of the root hair cell wall. Bar, 250 nm. (b) A rhizobial cell entrapped between closely adjacent plant cell walls as a result of root hair curling. Golgi bodies are evident in the cytoplasm. Bar, 250 nm. (c) Rhizobia between two adjacent root hair cells showing the fibrillar structure of the material surrounding the bacteria. Bar, 250 nm. (d) Rhizobia in the crook of a curled root hair showing an early stage of infection thread formation. The root hair cell wall has the appearance of having been cut away to allow the bacteria to gain access to the pocket formed by the developing infection thread. Bar, 500 nm. (e) Rectangular area from Figure 6.9 (d) at higher magnification, showing the termination of

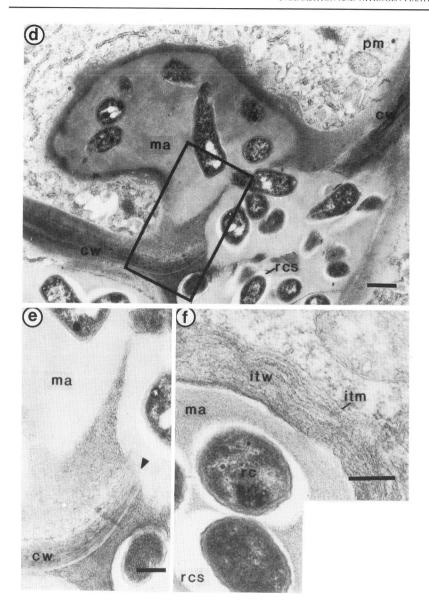

the cell-wall material (arrowhead) near the point of access of the rhizobia. Bar, 250 nm.
(f) Portion of an infection thread some way back from the tip showing the fibrillar material
in the infection thread wall arranged in a radial fashion around the matrix material. The
rhizobia within the matrix are themselves surrounded by an electron-translucent zone,
which may correspond to the bacterial capsular polysaccharide material. Bar, 250 nm.
cw, cell wall; f, fibrillar material; g, Golgi body; idm, infection droplet membrane; itm,
infection thread membrane; itw, infection thread wall; ma, matrix material; pm, plasma
membrane; rc, rhizobial cytoplasm; rcs, rhizobial capsule; rom, rhizobial envelope outer
membrane.

7.4 Graft unions

The process of graft union between stock and scion in grafting involves intimate contact between the cell walls of the two partners. Initially, the walls that are brought into contact by implantation of the scion form an extremely strong bond. This appears to be achieved by deposition of large quantities of pectin by both partners. This bond can form even in incompatible interactions, in which the scion will ultimately be rejected. In compatible interactions, the two partners produce callus tissue from the outer region of their stems, near the point of contact. The two calli grow together and, as they approach, nodules appear on the outer surfaces of the walls. Where the calli meet, the nodules flatten and coalesce, forming what appears to be a pectin-rich middle lamella. The pectin is then degraded in certain areas of the new wall, which become relatively thin. Plasmodesmata form in the thin-walled regions, and symplastic continuity between stock and scion is established.

The recognition process, which governs whether or not a graft will form between two partners, is not understood. It has been suggested that lectins may be involved but there is little evidence for this.

Summary

The cell wall provides protection against potential pathogens, both as a passive barrier and due to active deposition of wall material in response to infection. The wall is also involved in the triggering of other defence reactions through the formation of oligosaccharins. In the case of *Agrobacterium tumefaciens*, the cell wall aids infection by providing a specific point of attachment for the bacterium. Lignin and silica in some cell walls deter potential predators. The association between nitrogen-fixing microorganisms and higher plants to form nitrogen-fixing root nodules is initiated by interactions between the cell walls of the two partners. Graft unions between plants also involve complex interactions between the cell walls of the two partners.

References

Aldington, S. and Fry, S.C. (1994) Rhamnogalacturonan II – a biologically-active fragment. *J. Exp. Bot.*, **45**, 287–293.

Aldington, S., McDougall, G.J. and Fry, S.C. (1991) Structure–activity relationships of biologically-active oligosaccharides. *Plant Cell Env.*, **14**, 625–636.

Jacinta, T., Farmer, E.E. and Ryan, C.A. (1993) Purification of a potato leaf plasma membrane protein pp34, a protein phosphorylated in response to oligogalacturonan signals for defense and development. *Plant Physiol.*, **103**, 1393–1397.

Messiaen, J., Read, N.D., Van Cutsem, P. and Trewavas, A.J. (1993) Cell wall oligogalacturonides increase cytosolic free calcium in carrot protoplasts. *J. Cell Sci.*, **104**, 365–371.

Stacey, G., Greshoff, P.M. and Keen, N.T. (1992) Friends and foes: new insights into plant–microbe interactions. *Plant Cell*, **4**, 1173–1179.

Further reading

Albersheim, P. and Darvill, A.G. (1985) Oligosaccharins. *Sci. Am.*, **253**, (no. 3), 44–50.

Brewin, N.J. (1991) Development of the legume root nodule. *Ann. Rev. Cell Biol.*, **7**, 191–226.

Ryan, C.A. and Farmer, E.E. (1991) Oligosaccharins in plants: a current assessment. *Ann. Rev. Plant Physiol. Mol. Biol.*, **42**, 651–674.

Yeoman, M.M. (1984) Cellular recognition systems in grafting, in *Encyclopedia of Plant Physiology, New Series*, Springer, Berlin, Vol. 17, pp. 453–472.

8 The cell wall and reproduction

The complex process of sexual reproduction in higher plants involves a large number of specialized cell types, and many of these cells have distinctive cell walls. In this chapter, we follow the reproductive process from the formation of pollen grains through to the formation of seeds, highlighting the roles of specialized walls in these events.

8.1 Pollen mother cells

Pollen grains are derived from pollen mother cells, which arise within the anther. The walls of the pollen mother cells are similar to other primary walls when the cells begin their meiotic cycle, with numerous cytoplasmic connections between adjacent cells. However, as the formation of pollen by meiosis begins, the walls start to undergo a series of radical changes. First, the cell wall becomes heavily impregnated with callose, and the plasmodesmatal connections widen to around 1 μm in diameter, making up about 20% of the cell–cell interface. These broad connections bring about free cell–cell communication and may be involved in synchronizing the meiosis of neighbouring cells.

By the first metaphase of meiosis, the channels in the walls have closed up by further deposition of callose. This effectively isolates the cells from one another, though some apoplastic communication is still possible. This isolation may serve to permit the independent development of the pollen, which may be valuable since the pollen grains are now genetically distinct. As meiosis proceeds to the two-cell and four-cell stages, a particularly thick callose wall surrounds each dyad and then tetrad group, with thinner callose walls separating the cells within each dyad and tetrad group. As the tetrad stage proceeds, each pollen grain begins to lay down its own wall, and the callose wall is dissolved away by the action of β(1–3)glucanase (Figure 8.1).

8.2 Pollen grain walls

The pollen grain is surrounded by a wall of remarkable complexity. It is essentially a double wall, made up of an outer layer, the **exine**, and an inner layer, the **intine**. The exine may be further divided into the outer part, the **sexine**, and the inner part, the **nexine** (Figure 8.1b,c).

The two main layers (intine and exine) are quite distinct in their time of deposition, their structure and their chemistry. Following the normal sequence of cell wall development, it is the exine that is laid down first, deposition beginning when the grain is still enclosed in the callose wall following meiosis. The first stage of exine deposition consists of formation of a cellulose wall, called the **primexine**, within the callose layer. The principal features of the exine are detectable in the primexine, including rudimentary rods and surface spines. Once the microspore is released from the callose wall, it begins to grow, and the wall material that is laid down to accommodate growth is principally **sporopollenin**. By the time of pollen maturation, the exine is composed predominantly of sporopollenin. The site of synthesis of sporopollenin appears to be not the cytoplasm of the microspores, but the cells of the tapetum, which surround the pollen mother cells and their progeny.

Sporopollenin is a unique polymer, found nowhere else in nature. It is a polymer of carotenoid subunits (Figure 8.2), and is extremely resistant to degradation by both enzymes and chemicals, including strong acids. It is largely due to the high stability of sporopollenin that pollen grains can survive for very long periods, retaining their characteristic morphology and often their viability also.

The mature exine is often highly sculptured (Figure 8.3) but nevertheless conforms to a single basic design (Figure 8.1). The floor of the exine, the nexine, is a continuous layer, covering the entire pollen grain, except at a small number of apertures. From it arise radially orientated columns or rods, the **bacula** (singular: baculum). At their outer ends the bacula broaden to form either knobs, which may be partially fused to form intricate patterns, or a continuous roof, the **tectum**, which may be surmounted by spines or other ornamentation. The former type of surface is called a **pilate exine**, while the latter is called a **tectate exine**. In both types, the spaces between the bacula form elaborate chambers within the exine; in tectate exines, these chambers communicate with the outer surface through micropores in the tectum. The chambers contain glycoproteins and other diffusible substances important in pollen–stigma interactions.

The intine is laid down at a later stage than the exine, in contrast to which it is a relatively simple layer, though it may be quite thick. In composition it is similar to primary walls of somatic cells, containing cellulosic microfibrils and a matrix of hemicellulose, pectin and protein. It is exposed to the exterior at the apertures, and at these points it may be thickened and rather more complex in structure. Like the exine, it contains diffusible substances which are important in pollen–stigma interactions, and these substances gain access to the surface at the apertures.

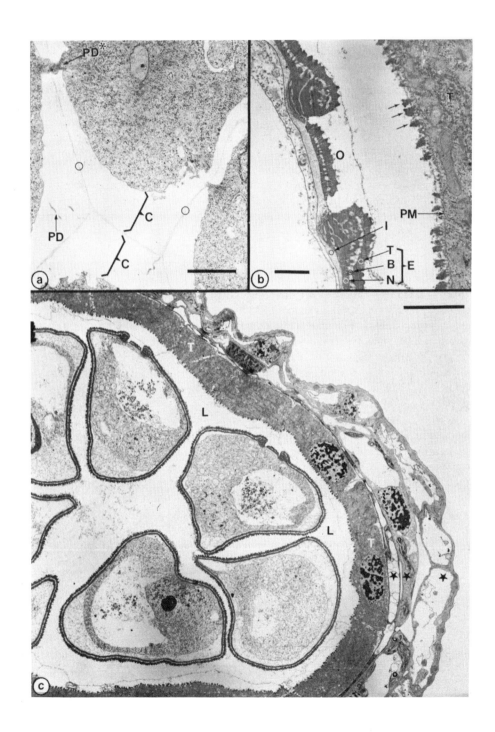

Figure 8.2 Example of the carotenoid monomers which, when polymerized, comprise the polymer sporopollenin.

The patterning on the outside of the pollen grains is extremely varied, and is characteristic of the species. Together with the extreme stability of the wall, this feature permits the classification and identification of pollens which have been embedded in anaerobic environments such as peat for thousands of years, and also of fossil pollen. As a result, pollen analysis gives the main clues as to the history of vegetation on earth (Box 8.1). It may also be used, indirectly, to provide evidence as to the history of animals, including humans, since pollen occurring in preserved excreta and in and around archaeological sites can be studied relatively easily.

Figure 8.1 (left) Development of pollen grains of *Avena sativa*. (a) Very early stage showing isolation of pollen mother cells by massive deposits of callose (C) which obliterates the primary walls. Callose is electron-transparent. Initially, the pollen mother cells are interconnected by cytoplasmic bridges (PD*), facilitating rapid distribution of nutrients and synchronizing the development of the cells. Remnants of true plasmodesmata (PD) are also present. Both types of intercellular connection are eventually sealed by callose deposition. Bar, 1 μm. (b) and (c) Developing pollen at later stage in which the callose has been digested away and the mature walls laid down; (b) is an enlarged (bar, 1 μm) detail from (c) (bar, 10 μm). In (c), maturing pollen grains are seen in the anther loculus (L), the wall of which consists of three layers of cells (stars) (the innermost layer partially crushed). The loculus is lined by a layer of nutritive cells, the tapetum (T). Pollen grain layers: intine (I), exine (E), nexine (N), bacula (B), tectum (T). Bacula and tectum together form the sexine. Oat pollen grains, like various others, have a single, circular pore where the exine bulges in a rim around a separate lid, or operculum (O), of sporopollenin sitting on a thickened pad of intine.

Figure 8.3   SEM of pollen grains of *Achillea millefolium* (L.) on the stigma of the same plant, showing their highly-sculptured exine surfaces. Bar, 20 µm.

Box 8.1 Pollen analysis in the study of the history of vegetation

Because sporopollenin is very stable, surface wall patterns on pollen grains can be recognized in pollen samples dating back to the first appearance of pollen-bearing plants in the Carboniferous Era, around 350 million years ago. However, the evolutionary changes that have occurred during that period mean that it becomes increasingly difficult in older samples to relate the pollen to particular species or genera. In the more recent strata, individual species or individual genera can be identified, depending on the degree to which the patterns differ between species in a genus.

Good preservation of pollen-grain structure only occurs in environments where both oxidation and mechanical agitation are avoided. For accurate dating of pollen samples, an undisturbed series of strata is necessary. Both requirements are met in waterlogged sediments such as those found in peat bogs and lake sediments. Such materials can be sampled by sinking an open-ended tube vertically downwards into the bog or sediment, and then removing it with its cylindrical contents, the 'core'. The individual layers can then be separated.

Analysis of the pollen begins with the removal of all other materials in the sample, by removing lignin by oxidation, polysaccharides by hydrolysis in sulphuric acid, and minerals by dissolution in hydrofluoric acid. The pollen are then identified by comparison with a reference collection, and the numbers of each pollen type are counted.

The simplest way of analysing the results is to calculate the percentage of total pollen represented by one particular species or genus, and to plot it against time. Taken together, all the graphs relating to one core make up the 'pollen diagram' for that particular core. Changes in the diagram occur at certain time-points, and probably correspond to changes in climate or to local changes in vegetation patterns. A series of diagrams deduced from cores taken from different locations can help to distinguish climatic rather than local changes.

More sophisticated analyses yield further information. With great care, absolute rather than relative pollen numbers can be calculated for a given volume of sample, and from this the number of pollen deposited per unit time can be estimated. This gives a better picture of the vegetation in terms of the numbers of individual plants in the vicinity of the sampling site.

The significance of patterning as far as the plant is concerned is not clear. The elaborate sculpturing of the exine permits the storage of diffusible chemicals, which are important in the pollen–stigma interactions that form the subject of the next section.

8.3 Pollen–stigma interactions

The surface of the stigma is generally moist, and the moistening of pollen grains landing on the stigma results in rapid diffusion of proteins and glycoproteins out of the pollen wall on to the stigma surface. These diffusing materials come both from the exine (from which they are released within seconds) and from the intine (from which they emerge in a matter of minutes, through the apertures). It seems likely that these components interact with materials on the stigma surface, including probably glycoproteins. These interactions may affect the adhesive qualities of the stigma, and hence the tenacity with which pollen sticks to the stigma surface. They may also affect the ease with which the pollen grains hydrate and germinate, and the ability of the pollen tube to grow down into the style.

All these steps can be barriers to effective germination when incompatible pollen grains are present on the stigma surface. Incompatibility may be due to the pollen being of a different species or genus, or it may be due to self-incompatibility. Foreign pollen tends to fail to hydrate or germinate; self-incompatibility, on the other hand, can be due to a failure to germinate, or a failure of the pollen tube to penetrate the stigma surface, to grow down through the style or to deliver the sperm successfully to the ovum. In the case of failure to penetrate the stigma surface, a layer of callose is often present in the walls of the stigma cells in contact with the pollen grain and this may present a mechanical barrier to penetration of the pollen tube.

The diffusion of materials from pollen wall to stigma surface is clearly involved in cell–cell recognition in these systems. This is clear from experiments in which mixtures of pollen are applied to stigmas. Normally incompatible pollen grains can be induced to fertilize flowers by application as a mixture with compatible pollen, the compatible pollen having been first sterilized by gamma irradiation to avoid competition. An even more effective method of overcoming incompatibility is to mix the incompatible pollen with diffusates from compatible pollen. Neither method, however, permits fertilization of a flower by all types of incompatible pollen, so clearly other mechanisms are also involved in incompatibility.

In *Brassica* species, the control of incompatibility resides in a single polyallelic locus, the S locus. Incompatibility is expressed when pollen and stigma have the same allele at this locus. Incompatibility is correlated with the appearance of large amounts of a particular glycoprotein in the cell walls of the stigma surface. Each allele is associated with a slightly different S-locus-specific glycoprotein (SLG). This family of glycoproteins are all of 55–65 kDa, with numerous N-linked oligosaccharide chains attached to the N-terminal half of the protein and a highly conserved cysteine-rich region in the C-terminal half. The genes from a number of these alleles have been cloned and sequenced, and their expression has been studied using the promotor region of the SLG gene fused to a β-glucuronidase reporter gene. In plants transformed with this chimaeric gene, gene expression is observed only in the pistil and the anther. A high level of expression is seen in the papillar cells of the stigma surface, correlating with the inhibition of pollen development at the stigma surface in incompatible interactions in *Brassica*. In anthers, a much lower level of expression is observed, and only in the tapetum during exine deposition. Thus the SLG found in pollen is clearly provided by the surrounding tapetal cells.

SLG is found to inhibit pollen tube growth *in vitro* but it is not clear what its role is in molecular terms. The only clue comes from a second gene found at the same locus, a highly conserved gene in which part of the sequence shows homology with the SLG gene and part is homologous to the catalytic domain of protein kinases. This may point to a signal transduction system involving protein phosphorylation.

In Solanaceae, a different self-incompatibility mechanism operates. The pistils contain abundant S-proteins, of 22–34 kDa. These extracellular glycoproteins are basic and N-glycosylated. The S-proteins from different plant lines show amino acid sequence conservation in five separate regions of the primary structure. Two of these conserved regions show strong sequence homology with the active sites of two fungal ribonucleases. The S-proteins also possess ribonuclease activity, and this activity may be part of the mechanism by which these proteins inhibit the growth of the pollen tube in the pistil in incompatible interactions.

In the Gramineae, inhibition of hydration of the pollen grains is an important step in the incompatibility reaction. In this family, the stigma surface is relatively dry and hydration involves transfer of water from within the stigma. Here also, a stigma glycoprotein appears to be involved in recognition and it probably interacts with pectin on the surface of the pollen exine.

8.4 The pollen tube

Once a pollen grain has germinated, a narrow tube – the **pollen tube** – grows down through the style and to the embryo sac containing the ova (Figure 8.4). The pollen tube must first penetrate the stigma surface and it secretes enzymes from its tip to accomplish this. Cutinase is frequently involved here. The pollen tube grows between the cells of the style, sometimes along a special channel. It is both guided and nourished by secretions from the stylar tissue. Its growth is unusual, in that it is by tip growth like a fungal hypha (Figures 8.5 and 8.6), rather than by extension of the longitudinal walls, which is the normal pattern for higher plants. Incompatibility can be expressed during tube growth through the style and may involve callose formation in the style.

The pollen tube wall is chemically distinct from somatic cell walls, since it is made up principally of callose ($\beta(1–3)$glucan) and $\alpha(1–3)$arabinan. Work carried out on *Nicotiana alata* indicates that the arabinan is present mainly in the outer part of what appears to be a two-layered wall. In this plant, the styles contain an arabinogalactan-protein which inhibits the growth of pollen tubes in incompatible interactions.

8.5 Cell wall food reserves in seeds

Once fertilization has occurred, seed development follows. Most seeds contain substantial storage reserves, to support the growth of the young

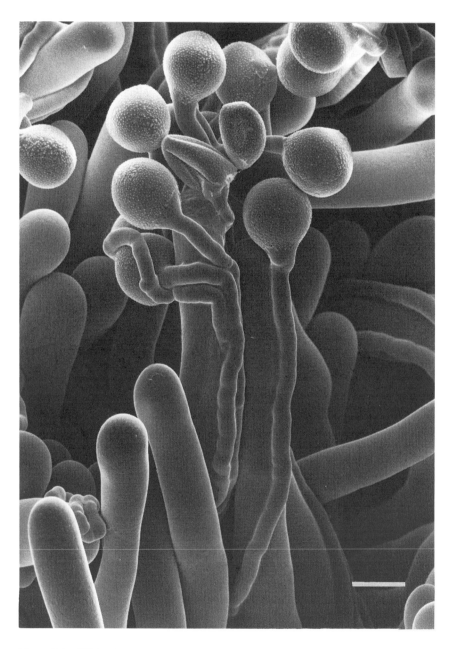

Figure 8.4   SEMs of germinating pollen grains of the opium poppy, *Papaver somniferum*. Several pollen tubes are shown growing over the surface of the papillae of the stigma; it also shows that on hydration, the furrowed pollen grains assume a more or less spherical shape. Bar, 50 μm. Note the external evidence of the callose plugs within the pollen tubes.

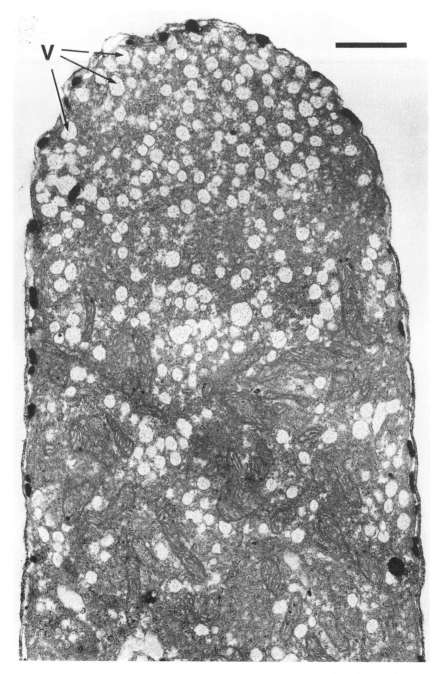

Figure 8.5   TEM of the pollen tube tip from *Tradescantia virginiana*. The picture shows the very thin wall at the tip, black lipid droplets close to the plasma membrane, secretory vesicles (V) at the tip and inside the tube, and a sub-apical mitochondrial zone. Bar, 1 μm.

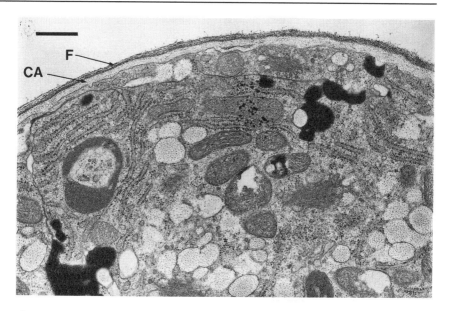

Figure 8.6   TEM of mature pollen tube of the same species as Figure 8.5 showing two cell-wall layers: an outer fibrillar layer (F) and an inner calloric layer (CA). Bar, 500 nm.

seedling until it can photosynthesize. The majority of these reserves are intracellular but in some instances the polysaccharides of the cell wall may act as food reserves. The nature and distribution of these cell wall reserves will be considered in this section; their degradation during germination will be discussed in Chapter 9.

8.5.1 Mannans and their derivatives

One major group of cell wall storage polysaccharides is based on a backbone containing predominantly β(1–4)-linked mannose residues. The group is found in the endosperms of a variety of plants.

'Pure' mannans, containing at least 90% mannose, are found in the endosperms of Palmae and Umbelliferae. In date and ivory nut, two types of mannan exist. Both are linear β(1–4)-mannans, but one type (Mannan A) has a degree of polymerization between 17 and 20, while Mannan B has a degree of polymerization around 80. Both contain 1–2% galactose and possibly some glucose. In ivory nut, *Phytelephas macrocarpa*, the mannan is crystalline and the endosperm is extremely hard.

Storage glucomannans are found in the endosperms of members of the Liliaceae and Iridaceae. They contain approximately equal amounts of glucose and mannose residues linked in a β(1–4) chain. A small proportion of galactose residues may be present as side-chains.

Galactomannans are found in the endosperms of all endospermic leguminous seeds, and also other species. The β(1–4)mannan backbone is substituted by many α(1–6)-linked galactose residues, with the ratio of galactose : mannose varying from 1 : 5 to 1 : 1. Galactomannan can constitute a very high percentage of the endosperm – 90% in the case of *Trigonella foenum-graecum* (fenugreek), and in this species the walls of the galactomannan-containing cells virtually fill the cells, which are dead at maturity. The numerous galactose side-chains prevent crystallization of the galactomannan and make it hydrophilic. As a result, the galactomannan takes up water when the seed is placed in a moist environment. If a hydrated seed is water-stressed, the water-retentive properties of the galactomannan act as a buffer and protect the embryo from desiccation (Figure 8.7). The galactomannan thus has a role both in water-retention and in nutrition.

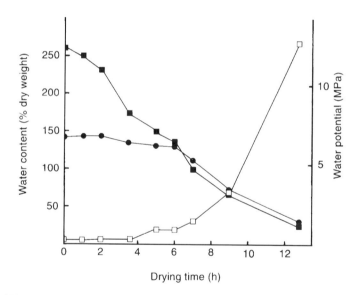

Figure 8.7   Water loss at 52% relative humidity from the endosperm/testa (■) and from the embryo (●) of the hydrated fenugreek seed in relation to the water-potential of the endosperm/testa (□). From Reid and Bewley (1979).

8.5.2 Xyloglucans

Xyloglucans are present in the cotyledons of many plants, including many of those legumes in which the reserve material is principally in the cotyledons rather than in the endosperm. Seed xyloglucans have traditionally been termed **amyloids** since, like amylose, they stain blue with iodine/potassium iodide reagent. However, they are based on a β-glucan backbone rather than an α-glucan one. The structure is similar to that of cell wall xyloglucans found in vegetative tissues, except that fucose is generally absent. The ratio of galactose : xylose : glucose varies between species and there is no evidence for any regularity of distribution of the side-chains along the backbone.

Like the galactomannans, the xyloglucans are hydrophilic. It is possible, therefore, that they also play a role in the water relations of the seed.

8.5.3 Galactans

These reserve materials are found in the genus *Lupinus*. They are probably not pure galactans but rather the galactan side-chains of rhamnogalacturonan I. The side-chains consist of linear β(1–4)galactan and are selectively lost from the wall during reserve mobilization, together with some arabinose from arabinan side-chains of RG I.

Summary

Pollen grains are generated within specialized callose walls and then form their own walls, which have characteristic chemical and morphological features. On the stigma surface, the pollen wall produces a secretion containing proteins and glycoproteins, which interact with stigma-surface components in recognition events. The pollen tube has a distinctive type of wall, made up principally of callose and α(1–3)arabinan. Once fertilization has occurred, certain developing seeds lay down large quantities of cell wall polysaccharides (mannans, xyloglucans and galactans) as food reserves.

Reference

Reid, J.S.G. and Bewley, J.D. (1979) Dual role for the endosperm and its galactomannan reserves in the germinative physiology of fenugreek (*Trigonella foenum-graecum* L.), an endospermic leguminous seed. *Planta*, **147**, 145–150.

Further reading

Kao, T. and Huang, S. (1994) Gametophytic self-incompatibility: a mechanism for self/nonself-discrimination during sexual reproduction. *Plant Physiol.*, **105**, 461–466.

Knox, R.B. (1979) *Pollen and Allergy*, Studies in Biology No. 107, Edward Arnold, London.

Meier, H. and Reid, J.S.G. (1984) Reserve polysaccharides other than starch in higher plants, in *Encyclopedia of Plant Physiology, New Series*, Springer, Berlin, Vol. 13A, pp. 418–471.

Reid, J.S.G. (1985) Structure and function in legume-seed polysaccharides, in *Biochemistry of Plant Cell Walls*, (eds C.T. Brett and J.R. Hillman), CUP, Cambridge, pp. 259–268.

9 Cell wall degradation and biotechnological applications

Cell wall degradation occurs in a wide variety of situations. First, it is important at certain points in the normal life of the plant, such as seed germination, xylem vessel formation, fruit ripening, abscission and perhaps growth (section 5.5). Secondly, it is an important part of the process by which other organisms degrade plant material, whether in the process of pathogenic attack by microorganisms or during digestion by herbivores. Thirdly, cell wall degradation is a key process underlying several industrial processes.

All these instances of cell wall degradation involve enzymic breakdown of cell wall polymers. Each of the enzymes involved carries out the lysis of a particular bond or bond type, and generally a battery of enzymes is needed in order for any significant wall breakdown to be achieved. The polysaccharide-degrading enzymes fall into the following broad classes.

(1) Hydrolases (EC 3.2.5)
(a) Endoglycanases These enzymes, also called endopolysaccharidases, hydrolyse internal bonds of polysaccharides.

(b) Exoglycanases These enzymes hydrolyse sugar residues from the end of a polysaccharide chain and are also called exopolysaccharidases. They may act as glycosidases, hydrolysing sugar residues attached to aglycones as well as to polysaccharides.

(2) Lyases, or transeliminases (EC 4.2.2) These enzymes cleave polysaccharides by elimination. They are known principally in pectin degradation.

(a) Endolyases, which cleave polysaccharides internally (Figure 9.1).

(b) Exolyases, which cleave single sugar residues from the non-reducing termini of polysaccharides by elimination.

9.1 Mobilization of food reserves in seeds

The cell wall storage polysaccharides were described in section 8.5.

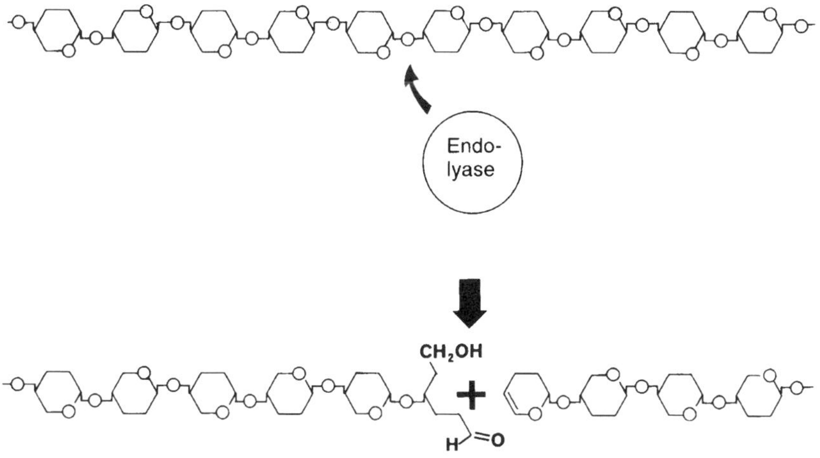

Figure 9.1 Diagram of endolyase activity.

They are mobilized during seedling germination and subsequent growth. They provide soluble carbohydrates for the young seedling, which can be used for the formation of other cell components and for respiration to provide energy.

The mannan group of reserve polysaccharides is present in the endosperm and must therefore be converted to a form which can be absorbed by the cotyledons. The pure mannans are degraded to mannose in date seed germination, which indicates that β-mannanase and β-mannosidase are probably involved. Those plants (such as *Asparagus officinalis*) which store glucomannan produce a β-glucosidase as well as a β-mannosidase in the endosperm. The mobilization of the galactomannans, which has been studied in some detail, occurs by the concerted action of three enzymes: an α-galactosidase, an endo-β-mannanase and a β-mannosidase. In fenugreek and guar, which have been studied extensively, the α-galactosidase and β-mannanase are synthesized *de novo* in the aleurone layer of the germinating seed and secreted into the galactomannan-containing cells of the endosperm. In guar, it has been found that the third enzyme, the β-mannosidase, is active in the ungerminated seed. The action of the three enzymes produces galactose and mannose, which diffuse from the endosperm into the embryo, where they are converted to sucrose and starch. In isolated endosperms from fenugreek, galactomannan breakdown occurs as in the intact seed, indicating that the embryo does not control the activity of the aleurone layer. Galactomannan breakdown is, however, inhibited by abscisic acid and stimulated slightly by ethylene and CO_2.

In lettuce seeds, a galactomannan cell wall storage polymer is present in the endosperm, though details of its structure are not known. During germination, the mannan is partially degraded in the endosperm by an endo-β-mannanase and an α-galactosidase, both produced by the endosperm. Complete breakdown of the resulting oligosaccharides to mannose is achieved by the action of a β-mannosidase which is produced by the cotyledons.

The mobilization of storage xyloglucans has been studied in detail in Nasturtium (*Tropaeolum majus*). Three enzymes are involved: an endo-β(1–4)glucanase, an α-xylosidase and a β-galactosidase. The endoglucanase is extremely specific for xyloglucan and is inactive towards substrates such as carboxymethyl-cellulose, which is commonly used to assay for endo-β(1–4)glucanases. It has endotransglycosylase activity as well as glucanase activity, and is very similar to the XET seen in vegetative tissue (section 5.5.2; Fanutti *et al.*, 1993). The α-xylosidase also shows specificity, since it does not hydrolyse *p*-nitrophenylxylose, which is the conventional substrate for assay of xylosidases.

Little is known about the hydrolysis of storage galactans. However, the fact that only galactose and arabinose are lost from the cell wall during germination indicates that galactosidases (or exogalactanases) and arabinosidases (or exoarabinases) may be involved. Exo-α-D-galactanase has been shown to liberate galactose from lupin cell walls (Figure 9.2).

In barley seeds, the major carbohydrate food reserve is a protoplastic one, starch, found in the endosperm, However, its mobilization involves aleurone cell wall degradation. The α-amylase which degrades the starch is produced in the aleurone layer, in response to gibberellin produced by the embryo. The aleurone cell walls are composed mainly of arabinoxylan, and gibberellin (and also ethylene) stimulate the production of a β(1–4)xylanase which degrades arabinoxylan. This is thought to facilitate release of α-amylase into the endosperm. Gibberellin also stimulates the production of β1,3-glucanase, which degrades the β1,3, β1,4 glucans in the endosperm cell walls.

9.2 Abscission

The shedding of plant parts occurs as a normal feature in the life of most plants and includes the loss of leaves, flowers and bark, as well as larger organs such as branches and roots. All these events involve the formation of one or more abscission zones, where the adhesion between cells that is mediated by the cell wall is overcome, bringing about cell separation.

Most of the work on the biochemistry and physiology of abscission has concentrated on the distal leaf abscission zone, which lies between

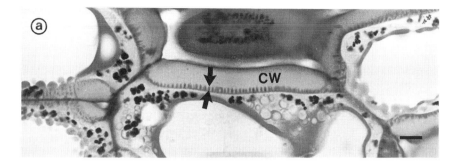

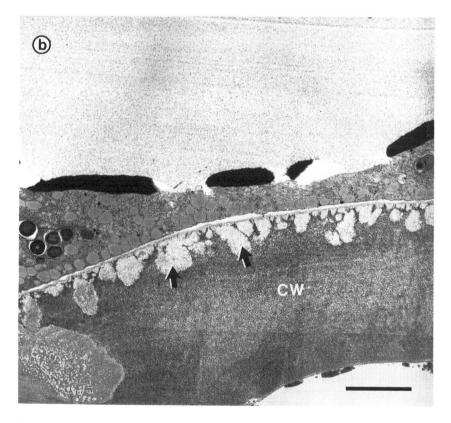

Figure 9.2   Micrographs showing the autolysis of cell-wall food storage components in lupin seed. (a) Light micrograph of a section stained with periodic acid-Schiffs reagent, showing up areas of the cell wall (CW) where enzymic digestion of the food store (galactan side-chains) is occurring (thick arrows). (b) TEM of the same tissue. The areas of the wall with lower electron density (thick arrows) are the areas of digestion. Bar, 2 μm.

the petiole and the pulvinus. Much of the work has been carried out on *Phaseolus* primary leaves and has often involved excised plant segments which include the abscission zone.

The process of leaf abscission may include degradation of the middle lamella at the separation zone, degradation of the whole cell wall in this region, and/or breakage of vascular tissue cell walls by mechanical forces resulting from cell enlargement on one side of the abscission zone. The degradation of the middle lamella involves pectin solubilization and degradation; degradation is probably brought about by a polygalacturonase, while solubilization may be due both to the degradation to smaller fragments and to an increase in pectin methylation, which results in breakage of calcium bridges between pectin molecules and an observable loss of calcium from the abscission zone.

Degradation of the whole wall involves cellulases as well as pectinases, and *Phaseolus* cellulase levels increase specifically in the cells closely associated with the abscission zone and proximal to it. The cellulase levels rise just before cell separation occurs, and only one of the four isozymes of cellulase in the tissue increases. Both cellulases and pectinases may be involved not only in weakening the cell walls at the separation layer but also in promoting wall plasticity and hence cell enlargement on the proximal side of the zone, resulting in breakage of vascular connections by mechanical shearing.

The abscission of leaves is promoted by ethylene, and ethylene is therefore likely to be involved in the induction of cell wall-degrading enzymes during abscission. Abscission is also influenced by auxin, abscisic acid, gibberellin and cytokinin.

9.3 Fruit ripening

Fruit ripening is often looked upon as the start of fruit senescence which, in addition to pigment synthesis and the production of volatiles, typically involves softening of the tissues. This is brought about by modifications in the structure of fruit cell walls. Ripening is a genetically programmed process which involves coordinated changes in a number of biochemical pathways. In certain 'climacteric' fruits, the ripening process is triggered by the autocatalytic production of ethylene (C_2H_4) to a threshold level. Examples include apple, pear, peach, plum, tomato, apricot, avocado, banana, mango, passion fruit and peach. Fruits which are not obviously climacteric include cherry, grape, strawberry and many citrus fruits.

The onset of ripening and the rate of tissue softening are of great interest both scientifically and commercially, having particular relevance to post-harvest quality and storage. Hence, there has been extensive research into the chemistry of ripening and biochemistry and genetic control of enzymes responsible for ethylene production and cell wall degradation.

9.3.1 Chemistry of cell wall degradation

Investigations into the chemistry of ripening have concentrated mainly on the changes in cell walls since these are the main determinants of texture and fruit firmness (section 10.1). It has been known for some time that softening of most fruits is accompanied by the dissolution of cell wall polymers, particularly those involved in cell adhesion (Figure 9.3), and that the polymers are often pectic in nature. Advances in cell wall research have enabled more detailed investigations to be undertaken. The use of modern methods of cell wall purification and extraction procedures (Chapter 2) have facilitated the detailed characterization of the carbohydrate composition of cell walls of several fruits and the changes that occur during ripening. Of particular importance has been the realization that some cell wall-degrading enzymes are quite difficult to inactivate by conventional procedures (e.g. short extractions in boiling ethanol). More vigorous treatments, including extraction in buffered phenol or phenol–acetic acid–water are necessary during cell wall preparation.

In addition to depolymerization and solubilization of polyuronides, ripening is usually accompanied by a decrease in the levels of cell wall neutral sugars, especially Ara*f* and Gal*p*. These sugars are usually metabolized. In some cases, cell wall uronic acid also decreases, although the GalA*p* released is not always metabolized.

Detailed fractionation studies of cell walls of several different fruits have shown that in many cases (e.g. apple, pear and kiwi fruit) ripening results in a relative increase in the levels of water and CDTA-soluble pectic polysaccharides. This is accompanied by a decrease in the levels of Na_2CO_3-soluble uronic acid and neutral pectic polymers, suggesting a possible function of these polysaccharides in cell–cell adhesion. The functional significance of the ripening-related decrease in pectic neutral sugars is not understood. An example of such changes in cell walls of pears is shown in Figure 9.4. In some fruits (e.g. tomato) ripening also involves a large decrease in levels of pectic polysaccharides that are closely associated with the cellulosic component of the cell wall.

9.3.2 Biochemistry of cell wall degradation

Cell wall degradation in ripening fruit results from the activity of a variety of enzymes. Their accumulation during ripening and their often con-

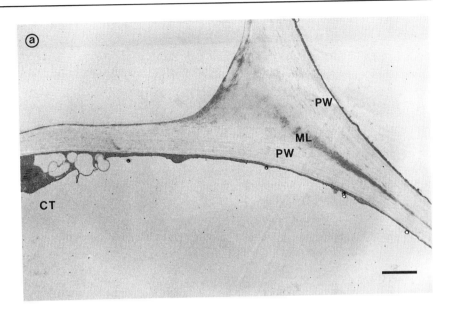

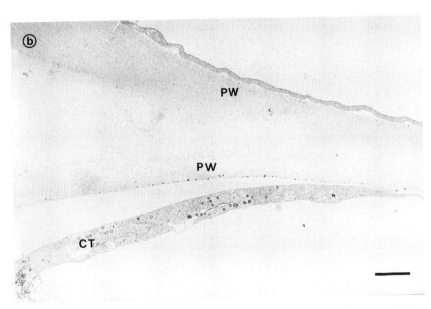

Figure 9.3 TEM of (a) unripe and (b) ripe tomato tissue. In (a), the middle lamella (ML) can be clearly seen between the primary cell walls (PW). In (b), the middle lamella has been degraded by enzymic activity and the cells are separating. CT, cytoplasm. Bar, 2 μm in each case.

a)

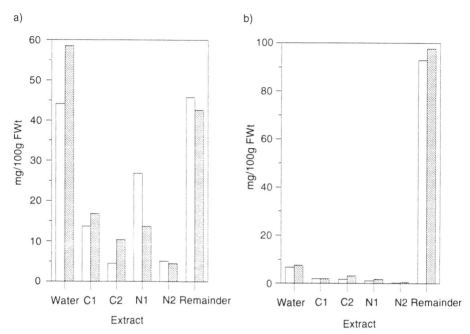

b)

Figure 9.4   Changes in the carbohydrate composition of pear (*Pyrus communis*) cell wall extracts during ripening. Cell wall material from unripe (white bars) and ripe (shaded bars) fruit has been subjected to sequential extraction as described in Boxes 2.5 and 2.6. (a) Total recovery of Ara + Gal + GalA (mg/100 g fresh weight). (b) Total recovery of Xyl + Man + Glc (mg/100 g fresh weight). The graphs show that ripening is accompanied by changes in the water solubility of pectic polysaccharides. C1, CDTA-1; C2, CDTA-2; N1, Na Co$_3$–1; N2, Na$_2$CO$_3$–2. See text for details.

certed activity will result, eventually, in degradation of many cell wall polymers leading to cell separation and associated softening (Figure 9.3). A number of enzymes have been identified which degrade a range of cell wall polymers. Several of these, including polygalacturonase (PG) and carboxymethylcellulase (CX-CMCase), reach high levels during ripening.

Enzymes with polygalacturonase activity hydrolyse α-(1–4)-linked GalpA linkages and include exoforms and endoforms. They should not be confused with bacterial pectate lyase which cleaves polygalacturonic acid by elimination (Figure 9.1). Some ripening fruits contain large quantities of exo-PG and barely detectable levels of endo-PG. However, in many such fruits pectin solubilization still occurs during ripening. Endo-PG is thought to be the main PG responsible for fruit softening: (a) there is often an inverse relationship between endo-PG levels and fruit firmness; (b) inhibition of endo-PG gene expression in tomato (see below) can significantly reduce softening; and (c) exogenously applied endo-PG can induce cell separation in fruit parenchyma and can mimic pectin degradation *in vitro*. Endo-PG has been detected in a wide range of ripening fruits but the majority of research has focused on ripening tomatoes.

Tomato PG comprises three isoenzymes: PG1, PG2A and PG2B. It is thought that PG2A and B isoenzymes are converted to PG1 by the addition of a 41 kDa ancillary subunit. This may be a regulatory step. The activity of tomato PG is also regulated by the degree of methyl-esterification of the pectin substrate, and it is most active in degrading demethylated polygalacturonic acid. Another cell wall degrading enzyme, pectin-methyl-esterase (PME) which demethylates pectins, has been implicated in regulating the susceptibility of polygalacturonic acid to PG. PME activity is found in several isoenzymes in tomato fruit, one of which increases during ripening.

Whilst Cx-cellulase enzymes can degrade CMC-cellulose, they are not able to degrade crystalline cellulose. It is likely that their role in ripening is one of degrading $\beta(1-4)$-linked glucans in the hemicellulosic fractions, such as mixed linkage glucans and xyloglucans. In avocado, Cx-cellulase activity increases to very high levels during ripening. It also increases, albeit to lower levels, during ripening of peach, tomato, strawberry and pear fruits.

A variety of other cell wall degrading enzymes are also found in ripening fruit including endo-$\beta(1-3)$-glucans, α and β galactosidases, endo β-$(1-4)$ mannanase, xylanase and a number of glycosidases. Some of these increase during ripening, indicating a role in the process.

The precise roles of many cell wall degrading enzymes during ripening have been studied in detail in several fruits. Changes in fruit firmness have been compared with changes in the levels of the enzymes, their activities, and the degradation of wall components. Further information has been obtained by measuring the levels of mRNA coding for the hydrolases and investigating changes in fruits which fail to soften, e.g. mutants of tomato plants. The most recent advance has involved the use of antisense technology in which plants may be genetically transformed with cloned cell wall hydrolase genes or antisense genes. By increasing or decreasing the activity of specific enzymes in a fruit, conclusions as to their function(s) may be drawn. Such investigations have highlighted the involvement of ethylene biosynthesis in initiation of ripening in climacteric fruits, and PG in the degradation of pectic polysaccharides during ripening of tomatoes. However, it is still not clear how this relates to fruit softening since softening occurs in some transgenic plants in which PG expression is all but inhibited, and softening fails to occur in certain transgenic tomato mutants in which enhanced PG activity causes the degradation of pectic polysaccharides.

9.4 Degradation by microorganisms

9.4.1 Pathogenesis

In the attack by microorganisms on living higher plant tissues, the major enzymes involved are those that attack pectin. Typically, the presence of pectic substances induces the formation of polygalacturonase and pectin lyase in both pathogenic fungi and pathogenic bacteria. These enzymes degrade the middle lamellae and primary walls of the host tissues. Often a number of isozymes are produced, especially in the case of the lyases. The enzymes are often subject to catabolite repression by glucose and other sugars, including the disaccharides and oligosaccharides that are typical products of pectin degradation. The action of these endo-degrading enzymes is often sufficient by itself to cause host tissue maceration and cell death. The cell death may be partly due to the osmotic sensitivity resulting from weakening of the walls, but it may also be due to the formation of pectin fragments which trigger cell death as part of the hypersensitive response (Chapter 7).

A number of higher plants have been shown to possess resistance to pectic enzymes. In some cases this is due to the presence of protein (probably glycoprotein) inhibitors of polygalacturonase.

Other enzymes are also involved in the pathogenic attack by fungi and bacteria. Some germinating fungal spores produce cutinase, which permits penetration of the hypha through the cuticle. Cutinase hydrolyses the primary alcohol esters of cutin and can also attack suberin. Small amounts of cutinase are thought to be present on the spore surface, and the cutin monomers produced on contact of the spore with cutin are thought to induce greatly increased production of cutinase.

Another enzyme that is sometimes formed by pathogenic fungi is endo-xylanase, which cleaves the $\beta(1-4)$ bonds of xylans. However, the significance of the enzyme in pathogenesis is unclear. Cellulases are also produced by pathogenic fungi and bacteria, especially those which rot wood.

9.4.2 Wood rot

Wood rot occurs in dead wood, in the heartwood of mature trees and in timber (Figure 9.5). It is characterized by the production of cellulase, ligninase or both. The ligninases are the subject of much current research, since their action is not well understood. They appear to be quite non-specific: they can degrade a wide range of aromatic and aliphatic C–C bonds. The reaction is an oxidation involving either hydrogen peroxide or hydroxyl radicals as the active agent. It is thought that the enzyme initiates a free-radical chain reaction, which causes extensive

Figure 9.5   A series of photographs illustrating the situations in which cell-wall degrading fungi can be found. (a) Timber decaying on the forest floor – a natural part of the 'carbon cycle'. (b) Dry rot in a London club caused by the entry of water after war damage. (c) An extensive outbreak of the dry rot fungus *Serpula lacrymans* in a damp cellar.

lignin degradation. There is some evidence that ligninase is induced in one wood-rot fungus, *Phaerochyte chrysosporum*, by the presence of lignin in the environment.

The cellulases concerned with wood rot are a complex of enzymes which act synergistically, involving both endo- and exo-β-glucanases. In the fungus *Trichoderma viride*, whose cellulase complex has been intensively studied, the endo-β(1–4)glucanase generates new non-reducing ends, from which cellobiohydrolase (an exoglucanase) cleaves successive cellobiose fragments. The cellobiose is then degraded by β-glucosidase. In certain fungi, cellulase action requires oxygen and the complex includes a component which catalyses cellulose oxidation.

9.4.3 Ruminants

Ruminants are capable of digesting cellulose, due to the presence of cellulolytic bacteria and sometimes fungi in the gastrointestinal tract. The degradation of cellulosic material is a result of the combined mechanical and enzymic activity of the system. The cellulolytic rumen bacteria,

(b)

Figure 9.5   *Continued*

chiefly *Ruminococcus* and *Bacteroides*, contain a complex of cellulolytic enzymes on their cell surface, as a result of which the plant cell wall is digested at those points at which the bacteria adhere. The complex includes endo-β-glucanases, exo-β-glucanases and β-glucosidases. Other rumen bacteria are able to degrade hemicelluloses and pectins, so most of the wall polysaccharides are digested in the rumen. In dry grass stems, the high levels of esterified phenolic acids (e.g. ferulic acid) in the walls can reduce the digestibility of the plant material.

9.5 Protoplast formation and use

Cell wall removal is an important step in certain processes by which genetic manipulation of plants may be achieved. Incubation of plant cells (generally suspension-cultured or leaf mesophyll cells) with a mixture of fungal cellulases and pectinases brings about the dissolution of the wall and the liberation of naked protoplasts. These protoplasts are extremely delicate, being sensitive to both osmotic and mechanical shock. They are also liable to be damaged by the toxic effects of some of the cell wall fragments produced during wall hydrolysis, since these fragments may actively induce cell death (section 7.1). They can, however, be cultured, and once the cell wall-degrading enzymes and cell wall fragments have been removed, they begin to form a new cell wall after a few hours. Once the wall has reformed, the cells can divide and grow normally in culture. This often means that the addition of appropriate plant growth substances to the medium can lead to cell differentiation, organogenesis and the regeneration of whole plants.

The period during which the wall is absent, or very thin, offers unique opportunities for studying the plasma membrane. For instance, patch-clamp techniques developed for animal cells can be used to study ion channels in the plant plasma membrane. This has recently brought about rapid advances in understanding the role of the plasma membrane in controlling ionic fluxes into and out of the plant cell (Wegner and Raschke, 1994).

During the period without a wall, it is possible to fuse protoplasts and produce viable hybrid cells. Fusion can be brought about by application of certain chemicals or by manipulating the electrical environment. The most commonly-used chemical fusion agent is poly(ethyleneglycol), usually known as PEG. Protoplast agglutination is caused by 10–50% PEG; subsequent dilution with a high-calcium, high-pH medium brings about fusion. Generally up to half the protoplasts can be induced to fuse by this mechanism. However, the procedure tends to be quite toxic, and it also tends to favour fusion between similar cells at the expense of hybrid formation. **Electrofusion** is now more commonly used; in this technique, an alternating electric field induces protoplast aggregation and the protoplasts can then be induced to fuse by a single high-voltage pulse of DC current.

It is sometimes possible to regenerate hybrid plants from these fusion products and hence achieve crosses which are not obtainable by conventional breeding techniques. Generally, however, hybrids can be regenerated into whole plants only when the two parental cell lines are fairly closely related, usually within the same genus. This limits the usefulness of the technique as a direct contributor to plant breeding programmes but it may nevertheless be an additional tool in the plant breeder's repertoire.

Foreign DNA may also be incorporated into protoplasts relatively easily. **Electroporation**, which uses manipulations of the electric field similar to those used for electrofusion, can be used to insert DNA into protoplasts, and microinjection is often easier in protoplasts than in walled cells. However, insertion of DNA into the cell (or even, by microinjection, into the nucleus) is not necessarily followed by stable incorporation into the genome, and difficulties in regenerating intact plants from protoplasts may mean that transformation using *Agrobacterium* is preferable. In those plants which are not infected by *Agrobacterium*, protoplasts are again an additional tool available to those who seek to produce transgenic plants of agricultural importance.

9.6 Technological applications

Cell wall degradation is of major importance not only in agriculture but also in a number of other industries. One important example is the brewing industry, in which cereal grain endosperm cell walls are degraded during the malting process. The degradation is carried out by the enzymes derived from the grain, involving proteases, endo-β-glucanases which degrade the β(1–3),β(1–4)glucans, exoglucanases, β-glucosidases, xylanases and pectinases.

The lignolytic enzymes are potentially extremely important in the utilization of wood waste such as sawdust. The presence of lignin in such material greatly retards the enzymic degradation of the carbohydrates of the wood to monosaccharides and disaccharides, which would be valuable feedstuffs for a variety of industries. Hence the development of stable enzyme systems capable of sustained lignolytic activity *in vitro* would provide major commercial benefits.

A major industry which already exploits cell wall degradation is the cultivation of edible fungi (Wood, 1989). The edible mushroom, *Agaricus bisporus*, is the main species involved but at least seven other species are now cultivated in significant quantities around the world. In the United Kingdom, mushroom production is the major off-farm use of surplus wheat straw, utilizing about 300 000 tonnes of straw each year. *Agaricus bisporus* produces a wide range of wall-degrading enzymes, of which the most abundant is laccase, a phenol oxidase which degrades lignin. Polysaccharidases secreted by *Agaricus* include cellulases, xylanases and glycosidases.

Summary

Cell wall degradation by enzymes occurs during the normal development of the plant, for instance in seed germination, abscission and fruit ripening. It also occurs during pathogenic attack and in ruminant digestion. Technological applications of cell wall degradation are important in brewing, recycling of wood waste, the cultivation of edible mushrooms and the production of protoplasts for genetic manipulation.

References

Fanutti, C., Gidley, M.J. and Reid, J.S.G. (1993) Action of a pure xyloglucan endoglycosyltransferase (formerly called xyloglucan-specific endo-β(1–4)D-glucanase) from the cotyledons of germinating nasturtium seeds. *Plant J.*, **3**, 691–700.

Wegner, L.H. and Raschke, K. (1994) Ion channels in the xylem parenchyma of barley roots. A procedure to isolate protoplasts from this tissue and a patch-clamp exploration of salt passageways into xylem vessels. *Plant Physiol.*, **105**, 799–813.

Wood, D. (1989) Mushroom biotechnology. *International Industrial Biotechnology*, **9**, 5–8.

Further reading

Brady, C.J. (1987) Fruit ripening. *Ann. Rev. Plant Physiol.*, **38**, 155–178.

Fincher, G.B. and Stone, B.A. (1981) Metabolism of non-cellulosic poly-saccharides, in *Encyclopedia of Plant Physiology, New Series*, Springer, Berlin, Vol. 13B, pp. 68–132.

Fischer, R.L. and Bennett, A.B. (1991) Role of cell wall hydrolases in fruit ripening. *Ann. Rev. Plant Physiol. Mol. Biol.*, **42**, 675–703.

Goodenough, P.W. (1986) A review of the role of ethylene in biochemical control of ripening in tomato fruit. *Plant Grow. Regul.*, **4**, 125–137.

Gray, J., Picton, S., Shabbeer, J. *et al.* (1992) Molecular biology of fruit ripening and its manipulation with antisense genes. *Plant Mol. Biol.*, **19**, 69–87.

Leng, R.A. (1973) Salient features of the digestion of pastures by rumi-nants and other herbivores, in *Chemistry and Biochemistry of Herbage*, Vol. 3, (eds G.W. Butler and R.W. Bailey), Academic Press, New York, pp. 82–129.

McDougall, G.J., Morrison, I.M., Stewart, D. *et al.* (1993) Plant fibre: botany, chemistry and processing for industrial use. *J. Sci. Food Agric.*, **62**, 1–20.

Stafford, A. and Warren, G. (1991) *Plant Cell and Tissue Culture*, Open University Press, Milton Keynes.

Waldron, K.W., Johnson, I.T. and Fenwick, G.R. (1993) *Food and Cancer Prevention: Chemical and Biological Aspects*, Royal Society of Chemistry, London.

Walton, J.D. (1994) Deconstruction of the plant cell wall. *Plant Physiol.*, **104**, 1113–1118.

10 Cell walls in diet and health

A significant part of the human diet consists of plant tissues (fresh or processed), hence cell walls will be ingested on a regular basis. The following sections give an indication of the impact that cell walls have on certain aspects of diet and health.

10.1 Cell walls as determinants of food mechanical properties (texture)

The textural properties of plant-based foods, particularly those containing intact cells, are strongly influenced by the mechanical properties of the cell walls. The force required to disrupt plant food materials during mastication will be determined, principally, by the ease of cell separation or cell breakage (Figure 10.1). Such properties will be affected by the state of maturity of the tissue, the degree of secondary thickening of the constituent cell walls, the interactions between different tissues and, importantly, the degree of processing. Tissues in which cell–cell adhesion is very strong may only be disrupted by breakage of the cell walls (Figure 10.1a). Such tissues are usually crunchy in texture and include uncooked vegetables and unsoftened fruits. Disruption of such tissues results in the release of cell contents. Tissues in which cell–cell adhesion is very weak may be disrupted through cleavage along the plane of the middle lamella (Figure 10.1b). Such tissues are usually soft in texture and include cooked (soft) vegetables and mealy fruits. Disruption of such tissues does not usually result in the release of the cell contents.

Edible organs of most fruits and vegetables are rich in immature parenchyma tissues. Cell separation is consequent to the depolymerization and/or dissolution of wall polymers bridging the middle lamella. This may follow biochemical events, as is the case in ripening fruits, or chemical changes, for example during cooking.

10.1.1 Tissue softening and cell separation during ripening

Ripening of edible fruits generally involves the softening of the tissues. In the early stages of ripening, disruption of the crunchy tissue involves

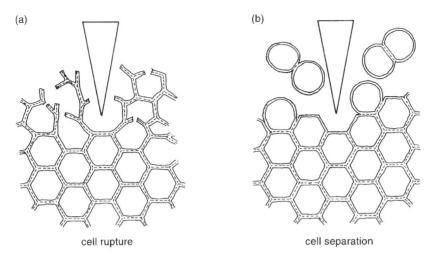

Figure 10.1   Tissue rupture by (a) cell rupture, typical of unripe fruits and uncooked vegetables (cell contents are released), and (b) cell separation, common to cooked vegetables and mealy fruits (cells remain intact).

cell rupture, although some cell separation may be evident (Figure 10.2a). However, in the later stages of ripening, particularly if fruits become mealy, disruption of the soft tissue involves much cell separation (Figure 10.2b). Cell separation is accompanied by, and probably results from, the depolymerization of the middle lamella pectic polysaccharides by cell wall degrading enzymes such as polygalacturonase, cellulase and pectin–methyl esterase (see section 9.3). Such changes appear to be reduced in the locality of lignified sclereids, as reported in pears (Figure 3.14).

10.1.2 Tissue softening and cell separation during heating

10.1.2.1 Cooking

Fresh vegetable tissues will rupture mainly by cell breakage (Figure 10.3). Heating vegetable tissues, e.g. during cooking, results in tissue softening due to an increase in cell separation. In some tissues (see below) initial softening will also occur due to loss of turgor. The mechanism underlying cell separation involves the heat-catalysed depolymerization of the pectic polymers of the middle lamella. The depolymerization involves β-elimination of the methyl-esterified polygalacturonic acid (Figure 10.4). However, chelation of calcium ions by organic acids released from the cells may also affect the ease of cell separation. Indeed potato and carrot cells can be separated not only by

(a)   (b)

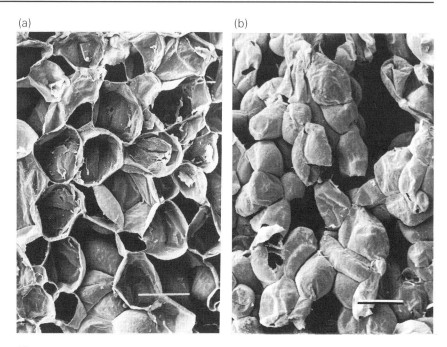

Figure 10.2   SEM of fracture surface. (a) A very crisp Granny Smith apple: most parenchyma cells have been broken open, releasing juice. (b) A very mealy Cox Orange Pippin apple: most parenchyma cells have remained intact, thereby retaining juice and giving a dry mouth-feel. Bars, 200 μm.

heat treatment (Figure 10.5a) but also by soaking the unprocessed tissues in CDTA (Fig 10.5b) (Chapter 2). Furthermore, the degree of heat-catalysed cell separation may be much reduced by the addition of Ca^{2+} to the cooking medium, even though Ca^{2+} can enhance the elimination. This suggests that the reduction of interpolymeric cross-links by β-eliminative cleavage may be more than compensated for by an increase in Ca^{2+}-mediated cross-linking. Intracellular starch, if in sufficient quantities, may promote cell separation during cooking. This results from the swelling of the granules which force the cell to round up.

10.1.2.2 Pre-cooking/blanching

During high temperature processes such as canning, the heat treatment required for sterilization can cause excessive softening of the fruit or vegetable. Oversoftening may be reduced not only by addition of Ca^{2+} to the canning media but also by pre-incubation of the living tissues at a lower temperature. For example, treating carrot root tissue at 50°C for 30 minutes will render the tissue extremely difficult to cook. The accepted hypothesis for this property is the heat-stimulated activity of

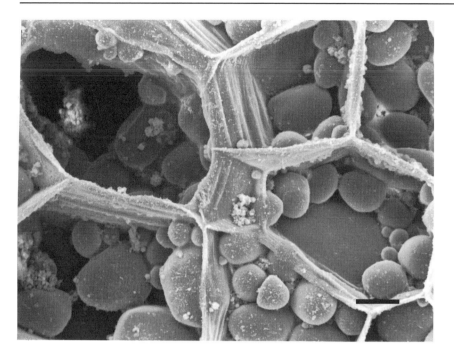

Figure 10.3   SEM of fracture surface (critical-point dried) of raw potato tissue. The line of fracture has involved cell rupture, revealing the intracellular starch granules. Bar, 50 μm.

pectin–methyl esterase (PME) in the cell walls. It is believed that this enzyme de-esterifies a proportion of the middle lamella pectic polymers. The demethylated pectic polymer exhibits a larger number of sites for cross-linking via Ca^{2+} bridges and will be far less prone to β-elimination during heat treatment.

10.1.3 Turgor pressure

In many tissues, particularly leafy ones such as lettuce, the texture will be affected by the mechanical properties of the cell wall and also the turgidity of the tissue. This will depend on both the elasticity of the cell wall and the osmotic potential of the intact protoplast (Chapter 5). Loss of water through evaporation or loss of membrane integrity (e.g. during heat treatment) will cause the tissue to become flaccid and to develop a less crunchy and somewhat rubbery texture.

10.1.4 Texture and maturation

In vegetables in which the edible organs are fast-growing immature tissues, maturation often results in toughening. This is due to the continued

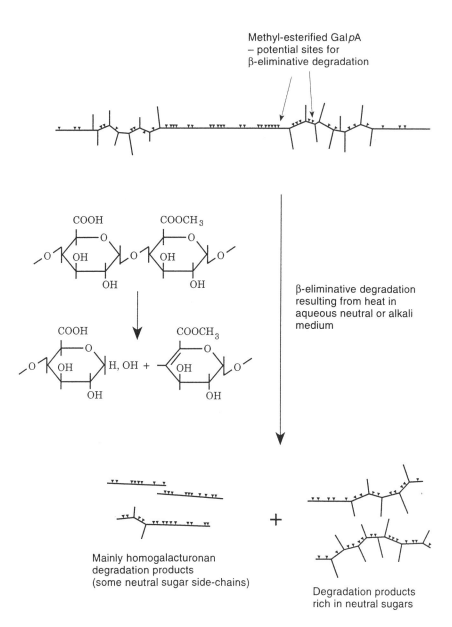

Figure 10.4   The β-eliminative cleavage of pectic polysaccharides during heat treatment. Such depolymerization is considerably enhanced by the presence of methyl-ester groups, and is probably a major factor in the solubilization of middle lamella polysaccharides during cooking.

(a)

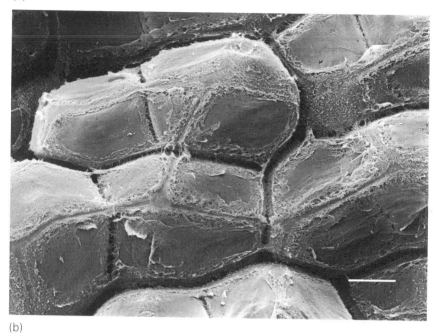

(b)

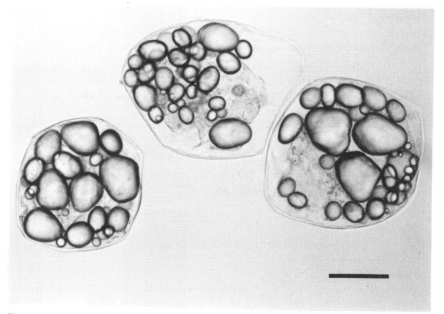

Figure 10.5 (a) SEM of fracture surface (critical-point dried) of cooked potato tissue. The line of fracture has followed the middle lamella. Bar, 50 μm. (b) LM of potato cells that have been separated by incubation of the tissues in a chelating agent, CDTA. Bar, 100 μm.

development of the tissues, particularly those in the early stages of secondary thickening (Figure 10.6). Asparagus stems provide a good example of maturation which continues in the stem tissue post-harvest. Toughening, particularly in the lower regions of the edible stem, has been shown to coincide with the synthesis of pectic–xylan–phenolic complexes (Waldron and Selvendran, 1992). These have been proposed as initials for lignification in the middle lamella region of the structurally important sclerenchyma tissues (Figure 10.6). Such cross-linking in the middle lamella reduces heat-induced softening by preventing cell separation.

A phenomenon which may be a maturation-related form of toughening involves the development of the hard-to-cook (HTC) defect in grain legumes, including those of the *Phaseolus* genus. Storage of such beans under conditions of high humidity and high temperature, prevalent in tropical countries, results in a decrease in the rate of softening during cooking due to an increase in cell adhesion (Figure 10.7). The extended cooking times (up to 10 hours) result in the excessive use of wood fuel and water.

In tissues which have undergone secondary thickening and lignification, cell adhesion will be extremely strong and will not be reduced significantly by either processing or physiological events. Furthermore, lignification will make cell rupture more difficult. Tissues which are rich in such cells (e.g. woody tissues) are not generally used as foods. However, some lignified tissues (e.g. cereal bran) may, after considerable disruption by milling, be incorporated into foodstuffs and provide an important source of dietary fibre (section 10.2).

Some non-lignified tissues, such as those of Chinese waterchestnut (CWC), fail to soften during extensive high-temperature treatments. For many years, researchers have compared the composition of their cell walls with those of comparable tissues, e.g. potato. The carbohydrate composition of the cell walls and the changes during heat treatment have not provided an obvious reason for the textural properties: the pectic polymers of CWC degrade in a similar fashion to those of potato. Recently, however, the cell walls of CWC have been found to be rich in phenolic acids: they fluoresce brightly under UV light in alkaline conditions (Figure 10.8). It has been proposed that these phenolics are responsible for cross-linking the polysaccharides of the adjacent cell walls, thus preventing cell separation during cooking (Parker and Waldron, 1995). Interestingly, experiments to induce cell separation by gentle removal of the bulk of the cell wall phenolics have revealed a curious fluorescent pattern in the cell wall which indicates a previously unidentified heterogeneity in the location or functional relationship of cell wall phenolics (Figure 10.9). This pattern appears to give a clear indication of the type and degree of cell packing in the water chestnut corm.

(a)

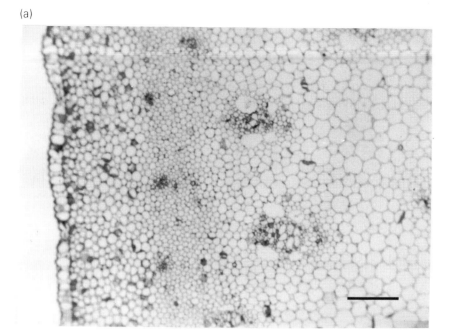

(b)

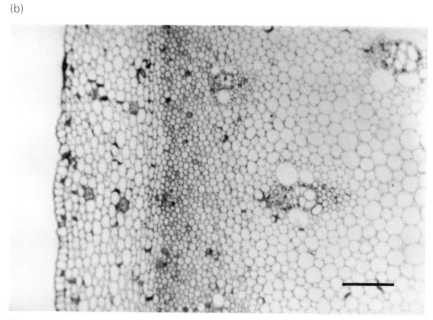

Figure 10.6   LM of asparagus spears: (a) edible region of the stem in which the sclerenchyma sheath is poorly developed; (b) tougher region of the stem in which the developed sclerenchyma has undergone some secondary thickening. Bars, 150 μm.

(a)

(b)

(c)

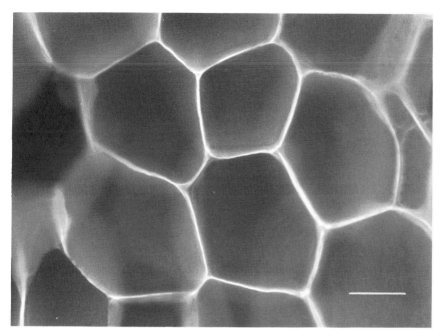

Figure 10.8   Fluorescence micrograph of the cell walls of Chinese water chestnut (CWC), pH 10. The yellow/green fluorescence indicates the presence of cinnamic acid derivatives such as ferulic acid, the cross-linking of which may impart the thermal stability for which CWC is well known. Bar, 40 μm.

10.2 Cell walls as a source of dietary fibre

The considerable interest in dietary fibre (DF) stems from the hypothesis that consumption of low-fibre diets is a common etiological factor in a large number of gastrointestinal and metabolic diseases of the developed world. It is generally accepted that a diet rich in foods that contain plant cell walls (e.g. fruits, vegetables, cereals) is protective against a number of diseases that are prevalent in Western countries, including constipation, diverticular disease, colorectal cancer, coronary heart disease, diabetes and obesity (Selvendran, 1991).

DF was initially defined as 'the skeletal remains of plant cell walls, in our diet, that are resistant to hydrolysis by the digestive enzymes of

Figure 10.7   SEM of the fracture surface of beans (*Phaseolus vulgaris* cv. Horsehead). (a) Recently harvested, after cooking for 30 minutes. Fracture has occurred by cell separation. Bar, 50 μm. (b) Hard-to-cook beans after cooking for 30 minutes. Fracture has involved rupture of the cell wall, revealing the cell contents. Bar, 25 μm. (c) Hard-to-cook beans after pressure cooking until soft. Fracture has occurred by cell separation. Bar, 25 μm.

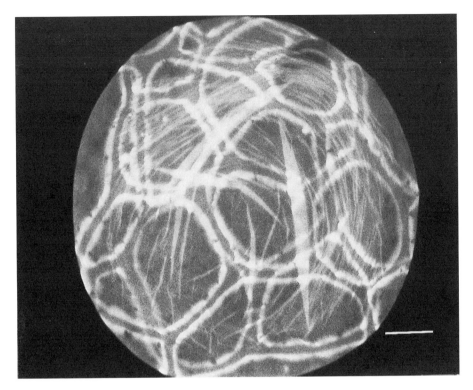

Figure 10.9　Fluorescence micrograph of the cell wall of a single CWC cell, separated from its neighbours by chemical treatments. The unusual fluorescence pattern remaining after cell separation suggests a degree of heterogeneity in the location/functional relationship of cell wall phenolics. Bar, 20 μm.

man'. This has since been extended to include 'all the polysaccharides and lignin in the diet that are not digested by the endogenous secretions of the human digestive tract', in order to include those polysaccharides present in some food additives, e.g. gums and thickening agents (Selvendran, 1991).

In most Western countries, the average daily intake of dietary fibre is approximately 20 g per person, much of which is derived from cereal sources. In contrast, a typical rural African diet is likely to contain 60–150 g DF per person per day (Selvendran, 1991).

10.2.1 Measurement of DF

For the purposes of quantification, DF is usually defined as dietary non-starch polysaccharide (NSP) and lignin. Of the variety of methods that have been developed, some measure DF gravimetrically after removing digestible food components such as starch and lipid, and measuring protein and ash content; others measure the carbohydrate components of a crude starch-free cell wall preparation. Methods differ in ease and cost and often give different values for a particular fibre source. Recently, the quantification of DF has been complicated further by the observation that some foods contain quantities of starch which is resistant to digestion and will, to all intents and purposes, behave like DF.

The value of total DF gives little indication of the physicochemical properties of the fibre, which will depend not only on the amount of cell wall (and other) polymers but also on their arrangement within the 'fibre matrix' and how that structure may be modified during processing. For example, wheat bran DF is more efficient as a laxative and faecal bulking agent than an equivalent quantity of apple DF. Furthermore, the beneficial properties of the bran may be lost if it is finely milled (section 10.2.3).

10.2.2 Physicochemical properties of DF

These will depend on several factors, including the structure and maturity of the cell types, the degree to which the plant material has been processed and the types of intracellular compounds. Any cooking or processing which affects cell wall integrity and/or cell–cell adhesion (section 10.1) will modulate the fibre properties by altering particle size, cell wall porosity and solubility of fibre components. These three parameters, combined with cell wall integrity, have a bearing on the ease of access of digestive secretions to the (nutritious) cell contents and, subsequently, diffusion of the digested material away from the fibre matrix. In addition, they will influence the water-holding capacity (WHC) of the fibre. The soluble fibre components, which include pectic polysaccharides, may, through their viscosity-enhancing and gel-forming properties, delay gastric emptying and possibly reduce absorption rates in the small intestine – a feature which is particularly important to diabetes. DF may also affect absorption rates of nutrients and other materials within the gastrointestinal tract by binding with them. For example, the non-esterified regions of pectic polysaccharides have the potential to bind cations, affecting their bioavailability. Furthermore, DF preparations from a variety of sources have been found to have the capacity to bind bile salts, affecting their re-absorption, and this may have implications in cholesterol metabolism.

10.2.3 DF in the human colon

Undigested food and other components will eventually pass from the small intestine into the colon. The majority of this material will consist of DF. However, in addition to polymers of cell wall origin, other endogenous NSP such as mucins and mucopolysaccharides and any undigested starch will also be present.

Once in the colon, polymers (particularly the polysaccharides) will be subject to degradation by the anaerobic bacterial flora. There are five predominant genera in the colon: *Bacteroides, Eubacteria, Bifidobacteria, Peptostreptococci* and *Fusobacteria*. The rate and extent of fermentation will be influenced by the accessibility of the polysaccharides to the bacteria and their enzymes. The main products of fermentation are short-chain fatty acids (SCFA) including acetate, propionate and butyrate; H_2; CO_2; and methane. The SCFA may be absorbed by the colonic mucosa and subsequently metabolized. Some of the gases will diffuse into the bloodstream and will be expired in the breath – the fermentation of fibre may be followed by monitoring breath hydrogen. However, the majority will be expelled as flatus.

The faecal mass, which is an important determinant of gut-transit time, will be determined by (a) the increase in bacterial biomass as a result of fermentation, (b) the quantity of unfermented fibre and (c) the water-holding capacity (WHC) of the fibre–bacteria complex. The extent of fermentation will be influenced by the quality of the DF. Fibre which is rich in primary cell walls that are low in lignin, as found in fruits and many vegetables, will be highly (>90%) degraded. This is due to the diffusion of the easily fermentable soluble fibre component out of the cell wall residue. This facilitates penetration of the fibre matrix by colonic bacteria, which can then degrade the non-lignified cellulose microfibrils. Such DF therefore stimulates bacterial growth and proliferation and leads to a moderate increase in faecal weight. In contrast, fibre which is rich in lignified cell types will be less fermentable. For example, only about 35% of wheat bran DF is degraded, due to its low solubility and high degree of lignification. The partially degraded bran fibre has a large WHC – the fibre–bacteria complex and associated water contribute to a large faecal weight, which is approximately twice that of vegetable fibre. Such a high WHC probably results from the curling of sheets of lignified periderm cells that remain after fermentation. The tubular structures so formed retain the aqueous phase through capillary action. Milling wheat bran to a fine particle size degrades this structure, reducing the water-holding properties even though, from an analytical view, the fibre has not changed. This highlights the importance of the physical and three-dimensional structural characteristics of DF in determining its physicochemical properties.

10.3 Bioactive cell wall and related components as protective factors in herbal products

During the past 30 years there has been considerable interest in the active components of some herbal drugs and medicines. Whilst much of the early research was inconclusive, usually due to a paucity of appropriate statistical treatment, there is now considerable evidence to suggest that some herbal remedies do have significant beneficial properties and that many polysaccharides of plant origin are commonly responsible for their bioactivities. Such polysaccharides are usually of high molecular weight (between 10^5 Da and 10^6 Da) and often soluble (hence their extraction in hot water infusions); and they may be associated with small amounts of protein. Many of the bioactivities so far identified appear to modulate aspects of the immune system. They include phagocytotic, anticoagulatory, hypoglycaemic, anti-inflammatory, anticomplementary, antibacterial, antiviral and, of particular interest, anti-tumour activity.

A large proportion of the bioactive polysaccharides are derived from fungal sources. Many, though, are found in eukaryotic plants, with examples from throughout the plant kingdom. Most are probably cell wall or plant gum in origin, with the structures typical of cell wall polysaccharides (Waldron and Selvendran, 1993), ranging from hemicellulose-like glucans to pectic polysaccharides. Whilst the modes of action of these polysaccharides are poorly understood, recent research (particularly into the activities of anti-tumour polysaccharides) has shed some light on the subject.

10.3.1 Anti-tumour polysaccharides

Much early work (pre-1980) on cell wall polymers exhibiting anti-tumour activity concentrated on fungal extracts, particularly (1–3)-linked glucans from basidiomycetes. More recently, cell wall-like polymers from other sources, particularly algae and higher plants, have been shown to exhibit such properties. These polymers include highly branched rhamnogalacturonan pectic polysaccharides from angiosperms, acidic polysaccharides and polyphenolic 'lignin-like' polymers from gymnosperms, and a selection of as yet uncharacterized algal and other cell wall components.

The modes of action of anti-tumour polysaccharides are poorly understood. Very few directly impair the mitogenic capacity of tumour cells; some have been reported to cause an increase in collagen fibre synthesis around the region of a tumour, thereby reducing cell proliferation. The most likely means by which anti-tumour polymers manifest their bioactivity appears to be via an interaction with cells of the immune system.

How can such orally administered/ingested anti-tumour polysaccharides come into contact with cells of the immune system? Rat-feeding studies using radio-labelled polysaccharides have shown that ingested anti-tumour polysaccharides, although undegraded by the enzymes of the alimentary tract, may be absorbed by some cells in the gastrointestinal epithelium, particularly those cells of the Peyer's patches which comprise a large proportion of the gastrointestinal lymphoid tissue. Absorption by such tissues will thus bring the polysaccharides into direct contact with cells of the immune system. This contact may be enhanced by their subsequent transport around the lymph system.

Several studies on fungal anti-tumour polysaccharides have suggested that they may act as T-cell adjuvants. T-cells are important in orchestrating the adaptive response (for further information, see Stites and Terr, 1991). More relevant to the anti-tumour polysaccharides from the cell walls of higher plants, a tentative model for the mode of action of one pectic anti-tumour polysaccharide has recently been put forward. In this model, the rhamnogalacturonan-I type polysaccharide is able to interact with receptors on the surface of both the tumour cell and certain cells of the immune system, such as lymphokyne-activated killer (LAK) cells and natural killer (NK) cells. By doing so, the polysaccharide is deemed to augment the cytotoxicity of these cells.

10.3.2 Anti-tumour polysaccharides in foods

Cell walls of edible plants are an abundant source of polysaccharides that contain residues and glycosidic linkages typically found in those polymers exhibiting anti-tumour activity. It has been suggested that some of the pectic polysaccharides that are solubilized from the cell walls of fruit and vegetables during cooking and processing may be of physiological significance post-ingestion (Waldron and Selvendran, 1993). The quantity of soluble RG-I type 'anti-tumour-like' pectic polysaccharides ingested by a 70 kg person consuming six typical 75 g portions of fruits and vegetables per day, is approximately 2 mg/kg body weight. This is only an order of magnitude less than the doses (on a mg/kg body weight basis) of RG-I type polymers found to be effective in significantly increasing the life expectancy of rats with a range of implanted tumours.

Perhaps these ingested polymers may be beneficial to health; it is generally accepted that a diet rich in fruit and vegetables is often coincidental with a low incidence of cancers, though this is usually attributed to the actions of dietary antioxidants. In addition to the soluble cell wall polysaccharides, some insoluble polymers may become solubilized during the fermentation process in the colon, the epithelium of which has an abundant covering of Peyer's patches. Indeed, the possible effect(s) of anti-tumour polysaccharides may provide another reason for some of the poorly defined beneficial properties of dietary fibre. What is in no doubt is that many cell wall-derived polysaccharides have the potential to act as immunomodulators.

Summary

A significant part of the human diet consists of plant tissues (fresh or processed), hence cell walls are likely to be ingested on a regular basis. The mechanical properties of cell walls, and how they change during processing and plant development, are important in determining the textural characteristics of plant-based foods. Cell walls also comprise a major component of dietary fibre, which is implicated in reducing a number of gastrointestinal and metabolic diseases of the developed world. Their potential immunoregulatory activity may have some bearing on this.

References

Parker, M.L. and Waldron, K.W. (1995) Texture of Chinese water chestnut: involvement of cell wall phenolics. *J. Sci. Food Agric.* (in press).

Selvendran, R.R. (1991) Dietary fibre, chemistry and properties, in *Encyclopedia of Human Biology*, Vol. 3, (ed. M. Yelles), Academic Press, London, pp. 35–45.

Stites, D.P and Terr, A.I. (1991) *Basic and Clinical Immunology*, Prentice Hall International, London, 870 pp.

Waldron, K.W. and Selvendran, R.R. (1992) Cell wall changes in immature asparagus stem tissue after excision. *Phytochem.*, **31**, 1931–1940.

Waldron, K.W. and Selvendran, R.R. (1993) Bioactive cell wall and related components from herbal products and edible plant organs as protective factors, in *Food and Cancer Prevention*, (eds K.W. Waldron, I.T. Johnson and G.R. Fenwick), Royal Society of Chemistry, Cambridge, pp. 307–326.

Further reading

Khan, A.A. and Vincent, F.V. (1993) Compressive stiffness and fracture properties of apple and potato parenchyma. *J.Text.Stud.*, **24**, 423–435.

Martin-Cabrejas, M.A., Waldron, K.W., Selvendran, R.R. *et al.* (1994) Ripening-related changes in the cell walls of Spanish pear (*Pyrus communis*). *Physiol. Plant.*, **91**, 671–679.

Mueller, E.A. and Anderer, F.A. (1990) Synergistic action of a plant rhamnogalacturonan enhancing antitumour cytotoxicity of human natural killer and lymphokine-activated killer cells: chemical specificity of target cell recognition. *Cancer Res.*, **50**, 3646–3651.

Southgate, D.A.T., Waldron, K.W., Johnson, I.T. and Fenwick, G.R. (eds) (1990) *Dietary Fibre: Chemical and Biological Aspects*, Royal Society of Chemistry, Cambridge, 386 pp.

Van-Buren, J.P. (1979) The chemistry of texture in fruits and vegetables. *J. Text. Stud.*, **10**, 1–23.

11 Outstanding problems for future research

The increase in interest in plant cell walls over the past few years has been due chiefly to two developments. First, improved methods of polysaccharide separation and analysis have opened up the structure of the wall to detailed study. Secondly, the realization that cell walls have a wide range of biological activities, especially the cell signalling activities of wall polysaccharides, has shown the importance of the detailed chemical structure of the wall and provided clues as to the reasons for its structural complexity.

Where will this research lead in the next few years? Certain lines of progress can be predicted with some confidence. For instance, the application of modern methods of analysis is likely to permit rapid clarification of the complex chemistry of the wall. The diversity of proteins present in the wall will be explored in detail, and their structure and function clarified. Immunocytochemistry will allow us to explore the detailed distribution of cell wall components within the wall. New spectroscopic techniques will be used to probe the structure of the intact wall. The degree to which oligosaccharins are significant in cell–cell communication will be clarified. The enzymic degradation of lignin will be thoroughly investigated.

There remain, however, many important areas of ignorance or technical difficulty in which progress is less certain. These offer great challenges to those interested in cell walls. First, the molecular mechanisms of the control of cell extension are still obscure as far as events like wall loosening are concerned. The nature of the growth-limiting bond(s) is unknown, though possible mechanisms by which bonds are broken and reformed during growth have been suggested. This is, perhaps, the most important unsolved problem, both scientifically and economically.

Secondly, the biochemistry of cellulose formation is still mysterious. The current hypothesis, that callose synthase and cellulose synthase are two forms of the same enzyme complex, offers hope but it is still not firmly established, even after much work. It is hoped that further detailed analysis of the polypeptides of the callose synthase complex and the corresponding genes, together with comparisons with the bacterial system, will yield the answer.

Thirdly, the question of how cell wall matrix components assemble round the newly synthesized microfibrils to form a new wall layer is only now being considered. Recent demonstrations of spontaneous self-assembly of wall components are very exciting. It seems likely that some degree of self-assembly occurs, even if the machinery of biosynthesis and the influence of pre-existing structures also play their part in controlling wall structure.

Fourthly, the commercial exploitation of recent breakthroughs in the enzymology of lignin degradation will not be easy. The potential rewards of developing commercial systems for lignin degradation based on lignin-degrading enzymes are considerable. However, the organisms which produce such enzymes are much too slow-growing and demanding in their conditions to be grown rapidly in culture. It will be necessary to transfer the relevant gene(s) to other microorganisms, and use these genetically engineered microorganisms either to degrade lignin directly or to produce sufficient quantities of lignin-degrading enzymes for commercial use after separation of the enzyme from the cells.

Fifthly, the mechanisms by which cell wall components act as signalling molecules in plants are likely to be hard to unravel. It has proved extremely difficult to identify receptor molecules and intracellular response mechanisms for the classical plant growth substances, some of which have been intensively studied for decades. It may be even harder to do this for cell wall signalling molecules, most of which operate in specialized physiological situations involving only a few cells, e.g. pollen–stigma and fungal spore–leaf surface interactions.

Finally, the interaction between plant cell wall polymers and human nutrition and health is likely to demand much attention in the near future. It will be hard to obtain definitive answers, both because of the complexity of the biological systems involved and because the enormous potential importance to medicine and industry tends, paradoxically, to get in the way of a dispassionate approach to the scientific problems involved.

Thus, there is no shortage of problems to be tackled in cell wall research. If this book has left the reader with an impression of our great ignorance about cell walls and how they work, it will have achieved one of its main aims.

Summary

Certain lines of research into plant cell walls are likely to yield much information in the next few years. However, a number of important questions are proving difficult to answer and continue to present researchers with major challenges.

Glossary

Below are given short explanations of terms which are not defined in the text. For words not included here, see the index for references in the text.

aglycone   Non-carbohydrate material in a molecule. Usually used in relation to glycosides, in which a monosaccharide is linked by a **glycosidic bond** to an aglycone.

aleurone layer   Outer layer of the **endosperm** of certain seeds. Contains protein bodies which store enzymes concerned with the breakdown of storage material in the endosperm.

allele   One of a number of forms of the same gene, all found at the same position on the chromosome.

anaphase   The stage of nuclear division at which sister chromatids (newly-formed chromosomes) separate and move away from one another.

angiosperm   Members of the Angiospermae, a subdivision of the Spermatophyta (seed plants). Distinguished from the **gymnosperms** by having enclosed ovules. Includes two classes, the **monocotyledonous** and **dicotyledonous plants**. Woody angiosperms are known as 'hardwoods'.

annular   Ring-like.

anther   In flowers, the part of the stamen which bears the pollen.

autoradiography   A technique in which radioactivity present in biological material may be visualized. Microscope sections are cut, a thin photographic film is applied to the section, and the radioactive emissions are allowed to darken the film. The film is developed, and viewed by light or electron microscopy. Dark silver grains are seen over radioactive areas of the section.

block copolymer   A macromolecule composed of chemically dissimilar segments connected in a linear sequence.

chelating agent   Substance which binds strongly to multivalent cations, by virtue of a number of anionic groups acting like pincers, e.g. EDTA (ethylene diamine tetraacetic acid).

coleoptile   Cylindrical sheath of leaf tissue, enclosing and protecting the young shoot in grass seedlings.

cotyledon Part of the embryo in seeds, acting either as a storage organ or in absorbing food reserves from the **endosperm**. **Dicotyledonous plants** have two cotyledons, and **monocotyledonous plants** have one.

cytosol The soluble phase of the cytoplasm, i.e. not including the membranes and membrane-bound organelles.

dicotyledonous plants (dicots) Plants whose seeds have two cotyledons. These plants form one of the two classes of **angiosperms**. Also known as broad-leaved plants.

differentiation The process by which cells change from the young, unspecialized state to the specialized state. Usually accompanied by structural changes that are detectable microscopically.

endomembrane system The internal membrane system of the cell, including the endoplasmic reticulum, Golgi apparatus and various small vesicles. Directed membrane flow occurs between these organelles and between them and other membranes (plasma membrane, tonoplast, etc.).

endosperm Seed tissue surrounding the embryo, containing food reserves.

epicotyl That part of the axis of a seedling lying between the **cotyledons** and the first node above the cotyledons.

glycoprotein Molecule containing both protein and carbohydrate; usually implies a higher percentage of protein than carbohydrate (cf. **proteoglycan**).

glycosidic linkages (or links or bonds) Linkage between the glycosidic hydroxyl (which is usually the hydroxyl on carbon 1) of a sugar and another hydroxyl (or occasionally sulphydryl or amino) group. In this book, the glycosidic bonds are described using an arrow or dash connecting the numbers of the relevant carbon atoms (e.g. $\alpha(1–4)$ for amylose), while the use of a comma between the numbers indicates the carbon atoms on one particular sugar residue which are linked to other sugars (e.g. 1,4). See Box 1.1 for examples of the use of this convention.

glycosyltransferase Enzyme which transfers a sugar residue from one molecule to another (cf. transglycosylase).

growth regulator Substance which, in very low concentrations, affects the rate of growth or pattern of **differentiation** of plant tissue. Also known as growth substance or hormone, though the term 'plant hormone' is now thought to be misleading.

growth substance See **growth regulator**.

gymnosperm Member of the subdivision, Gymnospermae, of the Spermatophyta (seed plants). Distinguished from **angiosperms** by having naked ovules, usually borne on cones. Includes conifers and similar plants. Woody gymnosperms are known as 'softwoods'.

haustorium   A branch of a fungal **hypha**, which penetrates the cell wall of higher plant cells and absorbs nutrients from the cytoplasm.

heteropolymer   Polymer containing more than one type of subunit; in the case of **polysaccharides**, the polymer (a heteropolysaccharide) contains more than one type of sugar residue.

HPLC   High-performance liquid chromatography; form of column chromatography in which a liquid medium passes over a solid phase, usually at high pressure.

hydrophobic force   (or bond)   Force arising from the tendency of hydrophobic molecules to aggregate in a hydrophilic environment, thus maximizing the number of hydrogen bonds between water molecules, and hence giving the lowest energy.

hypha   A filament of fungal tissue, which grows at its tip and is often branched. It is the main growth form of filamentous fungi.

hypocotyl   That part of the axis of a seedling lying between the **cotyledons** and the top of the root.

isozyme   One of a number of distinct proteins, all of which have the same enzyme activity. Also called 'isoenzymes'.

lumen   An internal space; e.g. the aqueous compartment within the endoplasmic reticulum or Golgi cisternae, and the interior of tracheids and fibres after the cytoplasm has degenerated.

maceration   The separation of the cells of a multicellular plant by mechanical, chemical or enzymic treatment.

metastable   Of a material, a physical state (e.g. a crystalline form) which is of higher energy than – and hence less thermodynamically stable than – another form of the same material, but whose transition to a more stable form requires a considerable energy input (i.e. has a high activation energy) and hence does not readily occur.

microfilament   Cytoplasmic filament composed of actin. Part of the cytoskeleton, its role in plants is uncertain, but it may include the control of organelle movement.

microtubule   Cytoplasmic tubule composed of the protein, tubulin. Part of the cytoskeleton, it is involved in the control of organelle movement, including chromosome movement and cytoplasmic streaming. It is also involved in the control of orientation of cell-wall microfibrils.

monocotyledonous plant   (monocot)   Plant whose seeds have one **cotyledon**. These plants form one of the two classes of **angiosperms**. Also known as 'narrow-leaved plants', they include grasses and cereals.

morphology   Form, shape or structure.

necrosis   Death, of cells, tissues or organs.

negative stain   In electron microscopy, a stain which accumulates around a structure, leaving the structure standing out as a light area on a dark background.

NMR   Nuclear magnetic resonance. A form of spectroscopy in which nuclei possessing nuclear spin (e.g. ^{1}H and ^{13}C) are induced to resonate in a strong, oscillating magnetic field. The resonance frequency depends on the chemical environment of the nuclei, and hence gives information about molecular structure.

non-reducing end (see Fig. 2.3)   In a **polysaccharide** or **oligosaccharide** chain, the end(s) which does not act as a reducing agent (i.e. lacks a free glycosidic hydroxyl). Branched chains have more than one non-reducing end.

oligogalacturonan   **Oligosaccharide** made up of galacturonic acid residues.

oligosaccharide   A carbohydrate containing 2–20 sugar residues (the upper limit is not well-defined).

oligosaccharin   **Oligosaccharide** which acts as a **growth regulator**.

osmoticum   A substance (e.g. sucrose, sorbitol or mannitol) which is added to a culture medium in order to raise its osmotic strength.

paracrystalline   Partially crystalline, with a significant degree of molecular order but without a fully crystalline lattice.

parenchyma   Thin-walled, highly vacuolated cells, forming the ground tissue of plants. Often photosynthetic; relatively unspecialized, sometimes capable of cell division.

petiole   Leaf-stalk, connecting the leaf blade or lamina to the branch or stem.

pollen tube   Tube by which nuclei pass from pollen grain to ovum during fertilization. The tube grows from the pollen grain into the **stigma**, and then down the **style** to the ovule.

polyallelic   A gene which may be present as any one of a number of different **alleles**.

polygalacturonan   Polymer consisting of a linear chain of galacturonic acid residues, usually $\alpha(1\text{–}4)$-linked.

polyprenol   A lipid alcohol, containing a sequence of up to at least 20 prenyl units, terminating in a primary alcohol. The prenyl unit has an isopentane carbon skeleton, which is usually unsaturated.

polysaccharide   Carbohydrate polymer consisting of at least 20 monosaccharides or monosaccharide derivatives.

proteoglycan   Polymer containing both carbohydrate and protein, in which the carbohydrate portion predominates (cf. **glycoprotein**).

protoplast   That part of the plant cell lying within the cell wall, i.e. the plasma membrane and all that lies within it. This term is often used to describe plant cells from which the cell wall has been removed.

proximal   Nearest the point of attachment to the plant body (opposite of 'distal').

pulvinus   Enlargement at the base of the **petiole** in certain plants.

radioimmunoassay   Assay method in which fixed amounts of antibody and radioactive antigen are mixed with unknown amounts of antigen in a sample under test. Radioactivity in the antibody–antigen precipitate is measured. The amount of antigen in the sample is calculated by comparison with appropriate standards.

Raman spectroscopy   A form of spectroscopy in which information is obtained from changes in wavelength that occur when radiation is scattered by certain molecules.

reducing end   The end of a **polysaccharide** or **oligosaccharide** chain which contains a free glycosidic hydroxyl, and thus acts as a reducing agent. There can only be one reducing end per molecule.

reticulate   Net-like.

scalariform   Ladder-like.

scion   In a graft, the portion of one plant which is grafted onto the **stock** of another plant.

screw axis   In a linear or helical polymer, the axis about which relative rotation of the subunits occurs moving from one subunit to the next. A 'two-fold' screw axis refers to a relative rotation of 180°, and a 'three-fold' axis to a relative rotation of 120°.

stigma   In the female part of a flower, the upper end of the pistil, usually enlarged to provide a surface for the reception of pollen.

stock   In a graft, the root system and part or all of the shoot of one plant, onto which is grafted a **scion** from another plant.

stoma (plural: **stomata**)   Small opening in the epidermis of leaves and some stems which opens to permit gas exchange and closes in conditions of water stress. Flanked by stomatal guard cells, which regulate the opening of the stoma.

style   In the female part of a flower, the main part of the pistil, connecting the **stigma** to the ovary.

synergism   Two or more factors acting co-operatively, so that their combined effects when acting together exceed the sum of their effects when each acts alone.

tensile strength   The ability to resist stretching, i.e. strength when under tension.

transglycosylase   An enzyme which transfers a mono, oligo, or poly-saccharide from one site of glycosidic attachment to another.

Young's modulus   The ratio of stress to strain, i.e. the amount of stress required to produce one unit of strain.

Index

DATE DUE